Geography in Bangladesh

This book provides an overview of the emergence of geography as a discipline in Bangladesh and the contributions made by local geographers towards the development of the country. It explores problems associated with population growth and poverty, landlessness and food security, land use and natural resource management, urbanism, climate change, disaster management and human health.

The volume shows how research and the study of geography in the 'periphery' can contribute in achieving progress in countries like Bangladesh and help them prepare against imminent disasters, ecological, social, economic shocks and uncertainties.

This book will be useful to students and researchers of geography, environment studies, disaster management, development studies, geo-informatics, geology, demography, sociology and South Asian studies with a particular focus on Bangladesh. It will also interest various policy makers and NGO professionals working in these and related fields.

Sheikh Tawhidul Islam is Professor at the Department of Geography and Environment, Jahangirnagar University, Dhaka, Bangladesh. He has been working in areas of environment, disaster management and climate change impacts for the last two decades. He has led more than 40 research projects and gained valuable experiences on environment and the development of Bangladesh and the region.

Alak Paul is Professor at the Department of Geography and Environmental Studies, University of Chittagong, Bangladesh. His research, focused mostly on marginalized people, uses qualitative methods and 'grounded theory approaches' in conducting research in the fields of human health and disease, environmental health and issues of man-made disaster events.

This volume is remarkable in bringing together 13 Bangladeshi geographers to write a geography of their country. Refreshingly empirical and unashamedly 'policy-relevant', it shows that there is a thriving and different geography beyond the Northern academy. Where better to be a geographer of the real world than Bangladesh?
   –**Professor Jonathan Rigg,** *School of Geographical Sciences, University of Bristol, Chair in Human Geography and former Director, Asia Research Institute (ARI) National University of Singapore*

Geography in Bangladesh
Concepts, Methods and Applications

Edited by Sheikh Tawhidul Islam
and Alak Paul

LONDON AND NEW YORK

First published 2019 by Routledge

2 Park Square, Milton Park, Abingdon, Oxon, OX14 4RN

605 Third Avenue, New York, NY 10017

Routledge is an imprint of the Taylor & Francis Group, an informa business

First issued in paperback 2020

Copyright © 2019 selection and editorial matter, Sheikh Tawhidul Islam and Alak Paul; individual chapters, the contributors

The right of Sheikh Tawhidul Islam and Alak Paul to be identified as the authors of the editorial material, and of the authors for their individual chapters, has been asserted in accordance with sections 77 and 78 of the Copyright, Designs and Patents Act 1988.

All rights reserved. No part of this book may be reprinted or reproduced or utilized in any form or by any electronic, mechanical, or other means, now known or hereafter invented, including photocopying and recording, or in any information storage or retrieval system, without permission in writing from the publishers.

Notice:
Product or corporate names may be trademarks or registered trademarks, and are used only for identification and explanation without intent to infringe.

British Library Cataloguing-in-Publication Data
A catalogue record for this book is available from the British Library

Library of Congress Cataloging-in-Publication Data
Names: Islam, Tawhidul, editor. | Paul, Alak, editor.
Title: Geography in Bangladesh : concepts, methods and applications / edited by Sheikh Tawhidul Islam and Alak Paul.
Description: Abingdon, Oxon ; New York, NY : Routledge, 2019. | Includes bibliographical references and index.
Subjects: LCSH: Geography—Research—Bangladesh. | Geography—Study and teaching—Bangladesh. | Geography—Bangladesh. | Geographers—Bangladesh. | Bangladesh—Geography. | Bangladesh—Social conditions. | Bangladesh—Environmental conditions.
Classification: LCC G76.5.B3 G46 2019 | DDC 910.72/05492—dc23
LC record available at https://lccn.loc.gov/2019001152

ISBN: 978-1-138-57061-0 (hbk)
ISBN: 978-0-367-73062-8 (pbk)

Typeset in Sabon
by Apex CoVantage, LLC

Dedicated to the humankind whose wisdom and prudential acts helped to bring us here.

Contents

List of figures	x
List of tables	xi
Notes on the contributors	xiii
Foreword	xv
Acknowledgements	xvii
List of abbreviations	xviii

1 Geography in Bangladesh: niche of the discipline
  in the context of global change   1
  SHEIKH TAWHIDUL ISLAM AND ALAK PAUL

PART I
Human geography   9

2 Urbanization in Bangladesh and urban research
  by geographers: a review   11
  NAZRUL ISLAM

3 Dimensions of gender geography in Bangladesh   26
  ROSIE MAJID AHSAN

4 Population geography in Bangladesh: in search of roots
  and approaches   44
  K. MAUDOOD ELAHI

5 Geographies of HIV/AIDS in Bangladesh: global
  perspectives, local contexts and research gaps   54
  ALAK PAUL

6 Geographical perspectives of scarcity and landlessness in rural households of Bangladesh 74
NASREEN AHMAD

PART II
Physical geography 95

7 Reconstruction of the palaeogeography and palaeoenvironment of Bangladesh: geographical perspectives 97
M. SHAHIDUL ISLAM

8 Environmental impact assessment in Bangladesh: applications of geographical tools and methods 124
DARA SHAMSUDDIN

9 Climate change crisis in Bangladesh 146
SHEIKH TAWHIDUL ISLAM

PART III
Applied geography (combining human and physical geography methods) 165

10 Geographical dimensions of food security in rural Bangladesh 167
MOHAMMAD ABU TAIYEB CHOWDHURY

11 Development induced displacement and resettlement practice in Bangladesh 185
HAFIZA KHATUN

12 Geographical methods in undertaking disaster and vulnerability research 202
SHITANGSU KUMAR PAUL

13 Natural resource appraisal in Bangladesh 222
MD. HUMAYUN KABIR

14 Literature reviews: searching for the deep roots
of geographical knowledge 241
NANDINI SANYAL

15 Local trends, the current state and the future
of geography 254
ALAK PAUL AND SHEIKH TAWHIDUL ISLAM

Index 260

Figures

4.1	The author with some of the leading population and human geographers of the post-1950s at the Symposium of the IGU Commission on Population Geography held in Zaria, Nigeria, 1978	47
6.1	Determinants of rural landlessness	91
7.1	The litholo-facies of each core at Boga lake and Pukurpara lake	102
7.2	Diatom records at Pukurpara lake	107
7.3	Holocene sea-level changes	114
7.4	A model showing a gradual vegetation succession from beach environment to inter-tidal zone followed by mangrove forest and, at last, marshy grassland	116
8.1	Impact assessment of the proposed interventions for two different types of impacts	143
9.1	Tropical cyclone category	152
9.2	Percentage of household by land ownership in Bangladesh	157
9.3	Daily wage rate in Taka	159
10.1	A conceptual framework of food security	169

Tables

6.1	The study sites	78
6.2	Number of households facing scarcity in the different (Bangla) months	80
6.3	Sources of current loans	85
6.4	Actual reasons for household landlessness over the floodplain	88
7.1	An example of sediment description after the Troels-Smith (1955) Scheme	104
7.2	Radiocarbon dates available from Bangladesh	110
7.3	Rates of changes of Holocene sea-level	113
8.1	Impacts of projects on Important Environmental Components (IECs) and consequential bearing on local people in flood/deltaic plains of Bangladesh	128
8.2	Procedure of the DoE (Bangladesh) for issuing environmental clearance certificate for different category of industries	131
8.3	Impacts of project components on different phases of implementation	133
8.4	Major environmental components of the study area against adaptation interventions	137
8.5	Strategic level impacts of adaptation interventions on IECs in climate change scenarios	139
9.1	Average maximum and minimum temperature fluctuations of selected weather stations in Bangladesh	148
9.2	Average monsoon (JJAS) rainfall compared to average annual total; highest and lowest deviation from monsoon average rainfall	150
9.3	Soil salinity in coastal Bangladesh	155
9.4	Financial loss as a result of disaster impacts in Bangladesh	156

xii *Tables*

9.5	Employed youth (15–29) by major occupation, weekly hours worked, sex	159
10.1	Key indicators of food insecurity	174
11.1	Types of losses eligible for compensation under the Development Partners' Policy and ARIPO of the GoB	189
11.2	A typical compensation package of a Development Partner funded project summary	193
12.1	Selected definitions of vulnerability in disaster discourse	203
13.1	Distribution of land type by inundation in Bangladesh	228
13.2	State of major forest types in Bangladesh	231
13.3	Renewable energy prospects in Bangladesh	234
13.4	Well-illuminated roof areas from Quickbird Image 2006 of Dhaka City	236
13.5	Solar PV based power generation potential in Dhaka	237
14.1	Necessity and purpose of conducting a literature review	244

Contributors

Nasreen Ahmad is Pro-Vice Chancellor and a former Professor in the Department of Geography and Environment at the University of Dhaka, Bangladesh. Professor Ahmad received her PhD degree for her dissertation on Landlessness in Bangladesh. She has authored several articles in reputed journals and has several book publications to her name.

Rosie Majid Ahsan is a former Professor in the Department of Geography and Environment at the University of Dhaka, Bangladesh. She received her MA degree from the London School of Economics, UK and a PhD degree from the University of Hawaii, USA. She has more than 35 research publications to her credit. She was also the Executive Secretary of the International Geographical Union (IGU) Commission on Commercial Activities.

Mohammad Abu Taiyeb Chowdhury is a Professor in the Department of Geography and Environmental Studies at the University of Chittagong, Bangladesh. He obtained his MA and PhD degrees in Geography from the University of Western Ontario (UWO), London, Canada. He is an active member of the Sustainable Agriculture Technology Network Asia (SATNET Asia) managed by UN-ESCAP.

K. Maudood Elahi is a former Professor in the Department of Geography and Environment at Jahangirnagar University, Dhaka, Bangladesh. He is currently the Pro-Vice Chancellor of Stamford University, Bangladesh and also a professor in the Department of Environmental Science at the same university.

Nazrul Islam is a former Professor in the Department of Geography and Environment at the University of Dhaka, Bangladesh and the Founder-Chair of the Centre for Urban Studies (CUS), Dhaka. He has also taught at the Asian Institute of Technology, Bangkok,

Thailand. He is an urban studies researcher and has authored several books and articles on urban development and related issues.

M. Shahidul Islam is a Professor in the Department of Geography and Environment at the University of Dhaka, Bangladesh. He received his PhD degree from the University of St. Andrews, UK. He has more than 30 research publications to his credit. Dr. Islam is a full member of the International Geographical Union's (IGU)'s Coastal Commission.

Md. Humayun Kabir is a Professor in the Department of Geography and Environment at the University of Dhaka, Bangladesh. He obtained his second MSc degree from the Asian Institute of Technology (AIT), Thailand, and received his PhD degree from Humboldt University in Berlin, Germany. Dr. Kabir has also published over 30 research articles and two books.

Hafiza Khatun is a Professor in the Department of Geography and Environment at the University of Dhaka, Bangladesh. She received her MA degree in Geography from the University of Alberta, Canada and her PhD degree from Dhaka University. Her research and teaching interests include social developments, dynamics and resettlement.

Shitangsu Kumar Paul is a Professor of Geography and Environmental Studies at the University of Rajshahi, Bangladesh. He received his PhD and MSc degrees from the Asian Institute of Technology (AIT), Thailand. He was also a fellow in the Department of Geography at the Norwegian University of Science and Technology, Norway.

Nandini Sanyal is a professional geographer. She received her MSc from the ITC, Netherlands, in remote sensing and an MA degree in Political Ecology from Durham University, UK. Her research merges satellite remote sensing with political ecology.

Dara Shamsuddin is a former Professor in the Department of Geography and Environment at Jahangirnagar University, Dhaka, Bangladesh. He was a US Fulbright Senior Scholar in the Department of Geography and Earth Science at the University of Wisconsin La Crosse, USA, where he worked on cumulative impact assessment methods of water sector projects.

Foreword

I welcomed my first Bangladeshi postgraduate student to Durham over 25 years ago and from then until my retirement I supervised to completion 17 research theses on the country of Bangladesh. Most of the researchers had established university posts at universities in Bangladesh and were taking a career break in order to pursue their interests full time. The Commonwealth Scholarship Commission provided financial support, along with the Christopher Moyes Memorial Foundation, the British Council and other funders. From my own point of view, this small industry of research on Bangladesh not only produced important academic results but also generated many continuing friendships. I greatly enjoyed this aspect of my career.

A generation of young geography researchers have returned from training in the Global North with their own ideas on how the discipline should be applied in Bangladesh. Many have immersed themselves in new technologies such as Geographical Information Systems (GIS) and remote sensing, which have the potential to generate and analyze vast databases of information, especially on environmental topics. Others have become convinced of the importance of new methodologies in human geography, for instance the qualitative techniques that are now yielding important insights into people's behaviour and decision-making, or, in physical geography, the application of wet-lab analysis. All of this amounts to a revolution in thinking about geography in Bangladesh.

It is important to note that the geography I am talking about here is not moving one-way, from Europe or North America. There are encouraging signs that journals of global significance and international book publishers are beginning to discover the wealth of talent in South Asia. The present book is just one example among many of the desire of Bangladeshi scholars to be heard and to contribute to their discipline in their own way. Readers will notice that an important element of this volume is applied geography. I take the message to be that human, physical and environmental geography have a vital role to play in understanding the many problems facing the country and in finding practical ways to solve them.

This book is a celebration of the many voices and approaches of geographers in Bangladesh. This diversity is a great strength and overall the book is proof that geography is both vibrant and relevant in this country.

Peter J Atkins
Em. Professor of Geography
University of Durham, United Kingdom

Acknowledgements

We are indebted to our mentors and teachers both at home and abroad who contributed in making our geographical mind. Firstly, both editors are very grateful to many British Geographers, especially the faculty members of the Department of Geography, University of Durham, UK, who have been with us over the decades. Secondly, we would like to show our respect to the contributors of this book who encouraged us throughout the process. Thirdly, special thanks are also due to the world-renowned publisher Routledge for facilitating us to publish this book; our special thanks to Shashank Shekhar Sinha of Taylor & Francis Group for showing confidence in our work and providing support to publish this volume. We are also thankful to Antara Ray Chaudhary for her continuous communication with us throughout the process. We are indebted to Chris Mathews from Apex CoVantage, LLC for providing critical comments which hugely contributed in the quality improvement of the book. The anonymous reviewers deserve special thanks for their critical appraisal of the manuscript; their comments and positive suggestions contributed hugely in making improvements. Lastly, we would like to express our gratitude to the many people who provided support, comments and assisted in the editing, proofreading and design. We should offer thanks to our students Nusrat Jahan Koley, Arifur Rahman Ovi, Munir Mahmud, Mohammad Mohaiminul Islam and Morsheduzaman who helped us in managing the materials of the book. We are thankful to the executives of Oriental Geographers (specially Maksudur Rahman) and the Journal of BNGA for their support. Above all, the first editor wishes to thank partner Nandini Sanyal and daughters Shoptorshi and Arundhuti for their support during this time. The second editor is also thankful to his wife Sumana Podder and both sons Anirudho and Arindom for their patience throughout the journey.

Abbreviations

AAG	Association of American Geographers
ADB	Asian Development Bank
AEZ	Agro-ecological Zone
AHP	Artificial Hierarchy Process
AIDS	Acquired Immunodeficiency Syndrome
APs	Affected Persons
ARIPO	Acquisition and Requisition of Immovable Property Ordinance 1982 II
BANBEIS	Bangladesh Bureau of Educational Information & Statistics
BARC	Bangladesh Agricultural Research Council
BARD	Bangladesh Academy for Rural Development
BBP	Bhairab Bridge Project
BBS	Bangladesh Bureau of Statistics
BGS	Bangladesh Geographical Society
BIEGIP	Bangladesh India Electrical Grid Interconnection Project
BMD	Bangladesh Meteorological Department
BNGA	Bangladesh National Geographical Association
BPDB	Bangladesh Power Development Board
BRAC	Bangladesh Rural Advancement Committee
CAPE	Convective Available Potential Energy
CARE	Cooperative for Assistance and Relief Everywhere
CBD	Central Business District
CC	Climate Change
CCL	Cash Compensation Law
CEGIS	Centre for Environmental and Geographic Information Services
CEIP	Coastal Embankment Improvement Project
CHTRDP	Chittagong Hill Tracts Rural Development Project

CHTs	Chittagong Hill Tracts
CIA	Cumulative Impact Assessment
CIDA	Canadian International Development Agency
CUS	Centre for Urban Studies\
CVI	Coastal Vulnerability Index
DCC	Dhaka City Corporation
DEEP	Dhaka Elevated Expressway Project
DEM	Digital Elevation Model
DFID	Department for International Development
DoE	Department of Environment
DP	Development Partners
DROP	Disaster Resilience of Place
ECAs	Ecologically Critical Areas
EFAPs	Erosion and Flood Affected Persons
EIA	Environment Impact Assessment
EMAP	Environment Management Action Plan
EM-DAT	Emergency Events Database
EMP	Environmental Management Plan
ERTS	Earth Resources Technology Satellite
FADEP	Integrated Food Assisted Development Programme
FAO	Food and Agriculture Organization
FAP	Flood Action Plan
FGD	Focus Group Discussion
FIVIMS	Food Insecurity and Vulnerability Information and Mapping Systems
FPCO	Flood Plan Coordination Organization
FSUP	Food Security for Ultra Poor
FSVGD	Food Security for Vulnerable Group Development
GAD	Gender and Development
GB	Grameen Bank
GBM	Ganges, Brahmaputra, Meghna
GDP	Gross Domestic Product
GED	General Educational Development
GHI	Global Hunger Index
GIS	Geographical Information System
GOB/GoB	Government of Bangladesh
GPS	Global Positioning System
HH	Household
HIV	Human Immunodeficiency Virus
HRD	Human Resource Development
ICCHL	Impacts of Climate Change on Human Lives
ID	Identification

xx Abbreviations

IDCOL	Infrastructure Development Company Ltd.
IECs	Important Environmental Components
IEE	Initial Environmental Examination
IFPRI	International Food Policy Research Institute
IGCP	International Gorilla Conservation Programme
IGU	International Geographical Union
ILO	International Labour Organization
IPCC	Intergovernmental Panel on Climate Change
IRR	Impoverishment Risk and Reconstruction
IUSSP	International Union for the Scientific Study of Population
JBARP	Jamuna Bridge Access Road Project
JICA	Japan International Cooperation Agency
JMBP	Jamuna Multipurpose Bridge Project
JMREMP	Jamuna-Meghna River Erosion Mitigation Project
KB	Krishi Bank
KII	Key Informant Interviews
KMHEP	Karnafuli Multipurpose Hydroelectric Project
LARP	Land Acquisition and Resettlement Plan
LDAZ	Local Diatom Assemblage Zones
LGED	Local Government Engineering Department
LGM	Last Glacial Maxima
LOS	Land Occupancy Survey
LPAZ	Local Pollen Assemblage Zones
LVI	Livelihood Vulnerability Index
MAT	Modern Analogue Technique
MCA	Multi-Criteria Analysis
MCE	Multi-Criteria Evaluation
MDGs	Millennium Development Goals
MHYVs	Modern High Yielding Varieties
MMCFD	Million Cubic Feet per Day
MV	Market Value
NEAM	North-East Asian Monsoon
NFP	National Food Policy
NGO	Non-government organization
NOC	Non Objection Certificate
NPRR	National Policy on Resettlement and Rehabilitation
NPWA	National Policy for Women's Advancement
NWAM	North-West Asian Monsoon
OBIA	Object Based Image Analysis
OSD	Occupation Skill Development
OWA	Ordered Weighted Average

PAR	Pressure and Release ModelRH Risk-Hazard
PMBP	Padma Multipurpose Bridge Project
PPP	Public-Private Partnership
PRA	Participatory Rural Appraisal
PRSP	Poverty Reduction Strategy Paper
PV	Photovoltaic
REP	Renewable Energy Policy
RG	Reconstruction Grant
RMP	Rural Maintenance Programme
RP	Resettlement Plan
RRMP	Regional Road Management Project
RS	Remote Sensing
RSs	Resettlement Sites
RV	Replacement Value
SAS	Statistical Analysis Systems
SD	Stamp Duty
SDG/SDGs	Sustainable Development Goals
SEA	Strategic Environmental Assessment
SEAM	South-East Asian Monsoon
SES	Socio-Economic Survey
SEVI	Spatial Multi-Criteria Social Vulnerability Index
SIA	Social Impact Assessment
SLR	Sea Level Rise
SoVI	Social Vulnerability Index
SPSS	Statistical Package for the Social Sciences
SRDI	Soil Resource Development Institute
SRNDP	South-West Road Network Development Project
SWAM	South-West Asian Monsoon\
TA	Technical Assistance
TBDLRP	Tongi-Bhairab Double Line Railway Project
TG	Transfer Grant
UDHR	Universal Declaration Human Rights
UK	United Kingdom
UNDP	United Nations Development Programme
UNFCCC	United Nations Framework Convention on Climate Change
UNFPA	United Nations Population Fund
UNICEF	United Nations International Children's Fund
UN-WFP	United Nation World Food Programme
USA	United States of America
USAID	United States Agency for International Development
VGD	Vulnerable Group Development

VGDUP	Vulnerable Group Development for Ultra Poor
WARPO	Water Resources Planning Organization
WB	World Bank
WCED	World Commission on Environment and Development
WED	Women Environment and Development
WFP	World Food Programme
WFS	World Food Summit
WID	Women in Development
WLC	Weighted Liner Combination

1 Geography in Bangladesh
Niche of the discipline in the context of global change

Sheikh Tawhidul Islam and Alak Paul

1. Introduction

The conceptual evolution of geography at a global level took place as a result of relentless introspective inquiry about 'what is geography' and related responsive actions taken by geographers at different times. The answer to this question had led western geographers to be strongly based on theoretical outfits which were primarily borrowed from Marxism (Harvey 1984), philosophy (Philo 1992) and social theory (Peet and Thrift 1989). But the geographers working in the Global South like Bangladesh did not have the same luxury or opportunity for various reasons to develop theoretical frameworks that fit the geographical problems of their region. The local geographers rather focused on empirical works on different physical, social, economic, ecological functions and processes of the land because of the pressing needs to understand the problems associated with those domains. Potter et al. (2004) also indicated that geography as a discipline has evolved, especially after the Second World War around a UK/Europe/North America focus (popularly termed as core focus) and paid little attention to the systems and processes happening in the Global South (known as geography in the periphery). They advocated for more works on different issues of the Global South and recommended a new vision of geography where both theories and empirical studies will contribute simultaneously (Ibid.). This book, against this backdrop, presents geographical findings produced by Bangladesh geographers to show how local geographers contributed to the development of Bangladesh through production of knowledge. The strengths they show in the works and the challenges identified will help to develop partnership projects between geographers working in the South and geographers based in the North and contribute to the overall advancement of the discipline.

2. Summary reflections of chapters

2.1 Human geography

Five chapters are included in the book under the human geography section. Urbanization process and urban research, gender dimensions of Bangladesh, evolutionary process of population geography, research gaps in health geographic research and rural poverty and landlessness in Bangladesh are discussed in different chapters. The authors show how geography as a discipline provided methods and tools in conducting investigations and doing useful analysis of the problems. Nazrul Islam, a noted urban geographer of Bangladesh, provides a detailed discussion on urban research conducted by geographers in Bangladesh. The chapter shows wide variations in the scope of research and methodology used by urban geographers in Bangladesh. Rosie M. Ahsan, a renowned gender geographer of Bangladesh, discusses the dimensions of gender geography in Bangladesh. She points out that without acknowledging women as a highly visible and powerful social and economic partners, economic and social strategy plans will fail to bring about progress. She showed how gender geography contributed in creating awareness among students, researchers and policy makers in Bangladesh in creating the social fabric. In searching for roots and approaches to population geography in Bangladesh, K. Maudood Elahi, an eminent population geographer, discusses the evolutionary process of the subject 'population geography' in the west and then traces its trends of teaching and research in Bangladesh. Finally, he elaborates on the contemporary research areas, challenges and future directions of population geography in the article. Alak Paul, a health geographer, mentions in his chapter that a significant number of medical or health geographers across the globe have given their attention to quantitative focus in HIV/AIDS diffusion across the regions but overlooked the qualitative approach. He argues that most of the research on HIV in Bangladesh has been performed by public health scientists and epidemiologists. Very little academic research has been conducted in Bangladesh and researchers have paid less attention to HIV that addresses the socio-economic or socio-geographic issues of the marginalized and stigmatized communities who are considered as the 'risk group' for HIV infection in Bangladesh. Nasreen Ahmad, a regional geographer, makes an attempt to understand landlessness as it operates in rural Bangladesh. Based on information on landless rural households within various agro-ecological zones of the floodplains of

Bangladesh, she shows the nature, processes and the causes of landlessness in Bangladesh.

2.2 Physical geography

Three chapters included in the physical geography section mainly focus on dynamic physical processes that contribute in configuring the lands of the country and also characterizing the natural resources base. Studies on quaternary environments, sea level rise and climate change aspects are also presented in relevant chapters of the book. M. Shahidul Islam, a renowned quaternary geographer of Bangladesh, highlights the reconstruction of palaeogeography and the palaeoenvironment of Bangladesh. He mentions that the archaeologist and geo-scientist can take the opportunity to infer about the palaeo-human occupancy in Bangladesh though a detail investigation of peat accumulation during the Holocene period. Dara Shamsuddin, a noted EIA expert of Bangladesh, writes his chapter on applications of geographical tools and methods in the field of environmental impact assessment in Bangladesh. He emphasizes the comprehensive technical skills of the EIA experts so that they can effectively synthesize both qualitative and quantitative data. He mentions that geography students throughout their undergraduate and postgraduate studies learn most of these techniques and are given opportunities to apply them in different research works, including environmental impact assessments. In the discussion on conceptual conundrums and challenges in coping with climate change crisis in Bangladesh, Sheikh Tawhidul Islam, a climate change expert, provides geographical explanation of how climate change aspects are conceptualized in Bangladesh, how programme activities are designed and implemented and, finally, identifies critical knowledge gaps in the area of climate change. He shows that the uncertainties and extreme events are the natural parts of the climatic systems of the country where seasons are characteristically distinct.

2.3 Contribution in environment and mixed methods

Six chapters are presented in the applied geography section. Authors in this section demonstrate how a fusion of human and physical geography methods may play useful roles in providing deep insights based on geographical investigations. Mohammad Abu Taiyeb Chowdhury, a noted development geographer of Bangladesh, discusses the theoretical/conceptual frameworks of food security concepts and the current state of food insecurity in Bangladesh from the context of sustainable development. He then describes the methodology employed in Bangladesh as a case study. He presents a summary of mapping, including the profile of highly food insecure regions. A detailed discussion of development induced displacement and resettlement practice in Bangladesh is provided in a chapter written by Hafiza Khatun, a renowned resettlement expert of Bangladesh. She evaluates existing land acquisition ordinance of Bangladesh government, assessed development partners requirements in relation to resettlement and rehabilitation practices. She presents 'good practices' for land acquisition, resettlement and rehabilitation for infrastructure projects in Bangladesh, with special attention to the Jamuna Multipurpose Bridge Project. A detailed discussion of geographical methods in undertaking disaster and vulnerability research is found in an article by Shitangsu Kumar Paul, a development geographer. Considering the multidimensionality of vulnerability, he reviews various definitions and models and presents a common geographical method of understanding vulnerability in disaster research. Md. Humayun Kabir, a natural resource expert of Bangladesh, in his chapter on natural resource appraisal in Bangladesh, argues about the urgent need for appraising existing natural resources (e.g., land, wetland, forest and energy) so that better planning and equitable sharing of natural resources can be done. Finally, Nandini Sanyal highlights the significance of literature reviews in conducting geographical investigations. The methods and strategies necessary to conduct a good literature review are discussed in her paper. She shows, by using some Bangladesh examples, how a literature review for geographical investigations can be done differently compared to other disciplinary experts. Finally, the last chapter depicts the scope of future geography based on the current strength of geographical knowledge and past geographical examinations.

3. Concluding remarks and the scope of the book

The two streams of geography, i.e., western geography popularly termed as 'core geography', which is based on theoretical threads, and geography of the South labelled as 'periphery geography', which focuses on empirical investigations (Potter 2001), generally remain in dichotomous positions in generating contemporary geographical knowledge. Effective actions to marry these two streams were not taken by geography schools or associations either from the Western part of the world (Potter et al. 2004) or from the periphery world located

Geography in Bangladesh 5

in the Global South. This dichotomous thinking was introduced by Western geographers (Hammett 2002), though their arguments were strongly criticized and opposed by others like Lawhon (2013); Lau and Pasquini (2008). Hammett (2002), in his works, undermines the works of third world geographers (pointing at South African geographers), as he mentions 'stereotypes of poor scholarship of the Global South academics' and 'failing to move beyond complaints'. Lawhon (2013), in this regard, strongly reacts and argues that the areas that Hammett (2002) sees as weakness are in reality the strengths of Southern/periphery geographers. Following this line of argument, the works of the geographers in Bangladesh included in this book may justify why and how empirical works could contribute to geographical knowledge productions as 'Southern geographical scholarship'. The works of Bangladesh geographers are mostly focusing on contextual settings; reading the change between people, place and time, focus is also placed on works that have a significant national need and contributes towards policy improvements. The review of geography articles (614 articles) published in two major journals from Bangladesh demonstrates the necessity and appropriateness of the works in relation to the need of the society and land. The results show that human geography has got the highest attention (49 percent contributions) from geographers in Bangladesh, followed by mixed-method geography (34 percent); the remaining 17 percent geographers contributed to physical geography areas. In the human geography sub-field, urban geography, population geography, rural and economic geography received the highest level of attention while ecology and biogeography, geomorphology and quaternary aspects were the main areas of contributions in physical geography. Mixed methods are used by the geographers working in agriculture, disaster and hazard research and natural resource management and planning. More than 90 percent of the works were empirically based investigations using statistical techniques (65 percent of cases), where field data was gathered to know the pattern of societal transformations and to read people, place and time more profoundly. Nearly half (43 percent) of the contributions focused on capital Dhaka city problems and adjoining central regions of the country. This account of the works published by Bangladesh geographers (representing the 'geographers of the periphery') indicates that they were not shy in undertaking exercises that are required for the society within the realm of debates on dichotomy (introduced by Hammett 2002) and related Western intellectual supremacy that arise out of the 'theory-based' versus 'empirical' styles of investigations. However, the contributions of geographers in different sub-fields of the discipline, given in the respective chapters, demonstrate in more detail how the problems have been conceptualized as contingent to the reality by applying geographical methods like 'spatial arrangements', 'association of activities' and 'geographical space'.

Bangladesh has experienced significant change during the last couple of decades; economic growth is steadily progressing (Bangladesh Planning Commission 2015), different social indicators are showing positive changes, successes are being achieved in physical, environmental and human induced hazards through capacity development of agencies at various tiers. These developments and/or changes continuously contribute in co-evolving new elements and processes and eventually the newly emerged elements are added to the existing set of issues. This fusion of new and old elements creates complex and dynamic systems/processes in the social order and also influences the natural environment in many ways. In these cases, new empirical investigations become necessary to shed light on these problems and help to see the whole problem domain relative to each other (old and new). On the other hand, the physical conditions of Bangladesh, i.e., flat topography, low elevation from mean sea level, unique hydrological conditions, the delta dynamics (land subsidence, sediment accretion, sea level rise) and interactions of all these processes with existing social elements/conditions create the need to undertake continuous investigations. The physical and human geographers in Bangladesh have been contributing towards that need since the beginning of the discipline in this country, in the 1950s. This book thus aims to demonstrate how geography in 'periphery', especially in Bangladesh, contributes towards sustained national development and to making the society resilient to shocks and uncertainties. This work may also support the call for 're-shaped vision of geography' (Potter et al. 2004), where both theory-based and empirical works will support each other for a stronger appearance of the discipline and eliminate the barrier that currently exists between 'core' and 'periphery' geographers.

References

Bangladesh Planning Commission. 2015. *MDG Bangladesh Progress Report*. Dhaka: Ministry of Planning.

Hammett, D. 2002. 'W(h)ither South African Human Geography?'. *Geoforum*, 43(5): 937–947.

Harvey, D. 1984. 'On the History and Present Condition of Geography: An Historical Materialist Manifesto'. *Professional Geographer*, 36: 1–10.

Lau, L. and M. Pasquini. 2008. '"Jack of All Trades?" The Negotiation of Interdisciplinarity Within Geography'. *Geoforum*, 39: 552–560.

Lawhon, M. 2013. 'Why I Want to Be a South African Geographer: A Response to Hammett's (2012) "W(h)ither South African Human Geography?"'. *Geoforum*, 47: A3–A5.

Peet, R. and N. Thrift. 1989. *New Models in Geography*. London: Unwin Hyman.

Philo, C. 1992. 'Foucault's Geography'. *Environment and Planning D: Society and Space*, 10: 137–161.

Potter, R. 2001. 'Geography and Development: Core and Periphery?'. *Area*, 33(4): 422–427.

Potter, R. B., T. Binns, J. A. Elliot and D. Smith. 2004. *Geographies of Development*, 2nd ed. London: Pearson.

Part I
Human geography

2 Urbanization in Bangladesh and urban research by geographers
A review

Nazrul Islam

1. Introduction

The need for urban research in Bangladesh was recognized early in the 1950s and such work started taking place with some earnestness in the following decades. Some of the research responded to specific planning requirements. Meanwhile, the urban population in the country has multiplied, the number of urban centres has gone up manifold, individual cities have expanded rapidly and urban problems have become much complex. The present chapter offers an overview of the status of urban research by geographers of Bangladesh especially since the 1970s. The chapter mainly refers to works by individual geographers but also to collaborative works, such as those at the Centre for Urban Studies (CUS), established in May 1972, led by a team of geographers, social scientists and urban planners. Before reviewing the record of urban research by geographers it may be relevant to present a brief resume of the patterns and processes of urbanization in Bangladesh.

2. Urbanization in Bangladesh

A BRIEF RESUME OF THE PATTERNS AND PROCESSES

Bangladesh is one of the world's most populous countries. It has a very poor land-human ratio; over 160 million people are crowded within only 147,000 square km, with a density of nearly 1,100 persons per square km. Even with such high average density, the country still remains overwhelmingly rural and agrarian. When the world population has become predominantly urbanized at 54 percent and even the Asian population as a whole is nearly 48 percent urban, the level of urbanization in Bangladesh is still quite low. In 2011, only 28 percent of the population of Bangladesh was urban. Despite such a low level of urbanization, the total urban population is quite large, being 43 million (BBS 2011).

Bangladesh has demonstrated a rather interesting trend of urbanization during the last five decades. It experienced very rapid urbanization during the 1950s and 1960s (the rate of urbanization being on average 5 percent annually). The rate of urban population growth in Bangladesh became even more rapid during the 1970s since, after the liberation of the country in 1971, the rate of urbanization was 7 percent, without reclassification of urban areas, and nearly 11 percent with reclassification. The growth rate had fallen only slightly in the 1980s, but still remains alarmingly high at more than 4 percent annually, which is more than three times higher than the rate of national population growth. The rapid rate of urban growth in Bangladesh is largely due to rural-urban migration, along with a moderate rate of natural increase and reclassification of urban boundaries and change of definition of urban areas. It has been estimated that the country would cross the 50 percent level of urbanization by 2045 (Mathur et al. 2013). Concentration of urban population in metropolitan areas has been a feature of urbanization in Bangladesh, as evident in the gross primacy of Dhaka and the growth of a few metropolitan cities like Chittagong, Khulna and Rajshahi.

2.1 Dhaka's increasing primacy

Dhaka, a historic city and the capital, is also the largest city in the country. It enjoys distinct primacy in the national urban hierarchy. In 1974, the city with a population of 1.77 million had over 28.3 percent of the nation's total urban population of 6.27 million. It was more than 1.99 times larger than the second largest city, Chittagong. In 1981, the primacy of Dhaka declined to some extent recording 25.7 percent of the national urban population. Dhaka's primacy was consolidated during the next decade. In 1991, the city contained 30.5 percent of the nation's 23 million urban population. The share increased to 34 percent in 2001 and was 33 percent in 2011, a persistently high primacy ratio. Dhaka became prominently larger than the second largest city between 1991 and 2011, from 2.91 times to 3.8 times its size.

Dhaka enjoys a domineering role in the country not only in terms of its share of population but also in terms of the concentration of civil and military administration, economy, trade and commerce (except port function). Educational, health, cultural and research activities are also highly concentrated in the capital city. Despite the government's policy of administrative decentralization in the early 1980s through the development of Upazilas (sub-districts) an overwhelming concentration of industries took place in the Dhaka megacity region during the last three decades. For instance, more than 75 percent of the nearly 4,000 export-oriented garment industries of Bangladesh are located in the Dhaka Metropolitan Region. In several other large industrial sectors, too, Dhaka has more than 80 percent of the national enterprises. At present, the national GDP of Bangladesh is estimated at about 235 billion US dollars (BBS 2016) and Dhaka city's contribution to the national GDP stands at 36 percent (Muzzini and Aparicio 2013: 2). Dhaka's increasing growth and primacy is partly explained by its historical background and location. The city is centrally located within the country and enjoys good accessibility with rail, road, water and air connections to all major towns and cities of the country. The evolution of Dhaka into a prime megacity has, however, created many serious problems for the city and its surrounding region. The physical environmental condition has deteriorated to an almost dangerous level, such as in the case of air, dust and noise pollution, river pollution, traffic jams, etc.

2.2 Increasing metropolization

The metropolitan cities with populations of more than one million also play important economic and socio-political roles in Bangladesh. There was only one metropolitan city, Dhaka, in Bangladesh in 1981; but in 2011, the number had increased to four, with Chittagong, Khulna and Rajshahi joining the ranks. These four metros together accounted for 63.87 percent of the total urban population in 2011, up from 35.69 percent in 1981. The dominance of business services, particularly finance and real estate services, is considerably higher in the four major cities relative to the rest of the country (GoB Planning Commission 2011).

Rather than having been stagnated, all the four metros have experienced higher population growths in the most recent decade (2001–2011), than in the previous one (1991–2001). The growth rates of the metropolitan cities have also been more than the urban population growth rates as a whole, underlining the fact that the scope of the processes of urbanization in Bangladesh are as yet unsaturated and that the country is expected to experience rapid urban growth in the coming decades, mainly fuelled by the growth in the metros. This process is expected to be led by Dhaka – the financial, cultural and business centre of the country – which has grown the fastest among the four metros, despite its large base.

2.3 Secondary cities and small towns

Below the megacity and metropolitan cities are the regional or secondary cities (with populations between 100,000 and 500,000), numbering about 38. These are mostly the old district towns/ cities. Some of these have a small industrial base but most of these are administrative, educational and commercial cities. Next in the level of urban hierarchy are the large numbers of small towns, with populations below 100,000 but above 20,000. These are generally the new district towns and upazila towns. They provide administrative, education, health and commercial services. A few have small industrial functions as well. At the bottom of the urban hierarchy are the numerous growth centres or market towns, including union centres, which provide social and commercial services to their immediate surroundings or villages. The intermediate cities and small towns serve as important centres of rural-urban mix and transformation of rural economy into non-farm or urban type economies. These urban centres have been transforming themselves with remittances from overseas migrants. The impact of overseas remittances is also evident in the rural landscape, with the increasing number of new pucca and semi-pucca houses.

3. Urban research in Bangladesh by geographers

Urban research in Bangladesh by geographers may be introduced as research on (a) pattern and process of urbanization at the national and regional levels (macro analysis), (b) research on the urban centre level, beginning with megacity Dhaka down to divisional/district/ Upazila centres, as indicated in the preceding section, (c) research on small areas or neighbourhoods, (d) research on thematic issues/problem issues, such as environment, economy, migration and demography, social structure, urban land and housing, urban services, urban poverty, urban management and governance, etc. The following review of urban research in Bangladesh by geographers focuses on the previous subject classifications but with a historical perspective, such as research before and after the liberation of Bangladesh in 1971.

4. Urban research during the pre-liberation period

The beginning of formal urban research in Bangladesh can possibly be traced to the early 1950s; it gained momentum in the 1960s. By the end of the 1960s, the base for higher education and academic research was broadened with the establishment of at least three general universities and a University of Engineering and Technology, where an Urban and Regional Planning Department started functioning since 1969. The geography departments at Dhaka University and Rajshahi University already had postgraduate teaching and research opportunities in the urban field. In addition, disciplines like sociology, social welfare and commerce also either taught aspects of urbanization or conducted research on it.

Professor Nafis Ahmad, the founding father of modem geography in Bangladesh, was the pioneer in urban research and publishing. He examined the urban pattern in Bangladesh (then East Pakistan) in the 1950s. Indeed, his paper entitled: 'The Urban Pattern in East Pakistan' was the first published journal article by a geographer in the country (Ahmad 1957). The paper also appeared in a modified form as a chapter in his famous book. *An Economic Geography of East Pakistan* published out of London by Oxford University Press in 1958, with a second edition in 1968 (Ahmad 1968). The paper discussed the size of towns, physiognomy of towns, classification of towns and the rural-urban movement of the population.

Led by Professor Ahmad, the geographers were the most involved and prolific urban researchers in the 1960s. Their research approaches were descriptive and empirical. Typical themes were urbanization, urban pattern, urban morphology, ecology or internal land use structure (Khan and Masood 1962), industrial location (Ahmad and Rahman 1962), peri-urban vegetable gardening residential pattern (Khan and Islam 1964) and population growth (Khan and Atiqullah 1965) or commercial structure (Majid 1970). Most of the urban research works and publications by geographers in the 1950s and 1960s were based on postgraduate (MA) theses and a few sponsored survey research works. When the first Physical (Urban and Regional) Planning Department was opened at the Bangladesh University of Engineering and Technology some geography graduates in the MA programme conducted urban research on regional planning (Mahmood 1971) or transportation, (Khan 1972). The theses were, however, completed after liberation. Bangladeshi geographers who had specialized on urban studies in universities abroad in the pre-independence period conducted research on British cities and only one on Dhaka (Majid 1970).

5. Urban research during the post-liberation period

The emergence of Bangladesh as a sovereign state in 1971 led to major changes in the economic and social life in the country. Social unrest, environmental hazards, famine and poverty, rural-urban migration were the pervasive issues. Initially, there was the declared policy of building a socialist economy under the leadership of Bangabandhu Sheikh Mujibur Rahman, Father of the Nation; but this policy was largely abandoned after his assassination in 1975 during a bloody military coup and change of government. The research paradigm on socio-economic development had also started to change after 1972. Likewise, urban research themes were also somewhat different from those in the earlier decades, although analysis of patterns and processes of urbanization, rural-urban relations and the distribution of urban centres by size remained a major theme from the beginning of the 1970s through the next decade. The attention of many geographers partly shifted to the study of urban poverty, slums and squatter settlements and marginal communities.

There has been geographical research interest in city size distribution. Bangladesh has shown an 'intermediate pattern' rather than a 'primacy pattern' like in some other third world countries (Elahi 1972). There has been fluctuation in the pattern of growth of urban centres with a few centres even experiencing decline. In the spatial distribution of the urban centres, there has been evidence of a 'random' rather than a 'uniform' or 'clustered' pattern (Islam 1975–76). There is also significant level of regional imbalance in the level of urbanization in the country, with the Dhaka Division (or central region) enjoying the highest level (Islam et al. 1991). In spite of the growth of a large number of urban centres in Bangladesh due to the absence of good physical communication facilities, more than one-third of the rural region of Bangladesh remained outside the field of daily urban influence of any major or even minor urban centre in the mid-1970s (Islam and Hossain 1976).

The proliferating slums and squatter settlements in the cities, particularly in Dhaka, received very high priority in the (unchartered) research agenda throughout the 1970s. The phenomenon of slums and squatters continued to remain a major research theme in Bangladesh deep into the 1980s and the 1990s and even into the first decade of the new millennium. Although slums had existed in the large cities and towns of Bangladesh for a long time, their presence became highly noticeable since 1972, after the end of the War of Liberation, which caused immeasurable damage to the national economy and triggered massive urban directed migration of the rural poor (Hossain 1972; CUS 1976). The population in slums and squatters together constituted between 10 percent to 50 percent of various city populations in the 1970s (CUS 1979). Within the city, the slums and squatters have been found to be located in a random manner in almost all parts of the city, but particularly in the already densely populated central areas and in marginal land along the rural – urban fringe (Islam and Mahbub 1991; CUS 1992). Squatters developed mainly on publicly owned land (CUS 1988). The physical environmental and housing conditions in slums and squatter settlements in Bangladesh cities, particularly Dhaka, have remained deplorable (CUS 1988). There has also been mixed impacts, with some slum dwellers experiencing upward mobility while others have remained either stationery in their socio-economic position or have even moved downward (CUS 1990). The Centre for Urban Studies also felt the need to know more about the socio-economic mobility trend (CUS 1990). Social mobility of poor female migrants in Dhaka was the subject of an in-depth study by geographer Shanaj Huq-Hussain for her doctoral thesis at the School of African and Asian Studies in London (Huq-Hussain 1992). Several studies related to policy on slums and squatters have recommended accommodation, rather than dislocation, of such communities since they largely solve their housing problems on their own (Hasnath 1977; CUS 1990; Islam et al. 1991).

Following the 1971 Liberation War and the 1974 flood and famine, rural push factors became irresistible and rural-urban migration was inevitable. The magnitude of the problem and its dynamics received the natural attention of geographers (CUS 1979, 1981; Khan 1982). Other streams of migration and internal urban mobility also received research emphasis (Ahsan and Sinowar 1984; Mahbub 1986). Several of these studies analyze the migration behaviour of women (Begum 1979). Various population and demographic characteristics of the urban population have also been studied for Dhaka, Khulna and other cities (Islam 1986; Saleheen 1992). Rural-urban migration has shown persistence in the flow towards the large cities, more particularly the largest and prime city, Dhaka. Both rural push and urban pull factors have played their role. However, the economic factors and natural hazards have been the major causes of rural push (Mahbub and Khatun 1990; Islam and Mahbub 1991). Distance as an intervening variable was also found to be a critical element, most of the migrations having taken place from the districts closest to the city of destination. Kinship network was effective in creating chains in migration flow and in adjustments at the urban end (Islam 1986, 1990). Urban adaptation by migrant poor women was studied by Huq-Hussain (1992).

The whole phenomenon of urban poverty and the problems of the urban poor became a policy issue in the 1970s and needed serious programmes to be taken for action. The Government and UNICEF decided to initiate some programmes. This needed adequate information and knowledge and the Centre for Urban Studies, under the leadership of geographer Nazrul Islam, was invited to conduct a comprehensive study on the urban poor in Bangladesh (CUS 1979). Poverty continued to remain pervasive in the following decades and so has the need for research on various aspects of poverty (CUS 1990; Islam et al. 1996a). But research on approaches to urban poverty alleviation is yet to be seriously undertaken.

The CUS 1979 Urban Poor Study (CUS 1979), based on field surveys in four large cities, the CUS 1990 Urban Poor Study (CUS 1990) and the 1995 (Islam, 1996b) study based on secondary sources and field surveys covering all important aspects of urbanization and urban poverty offer fairly comprehensive understanding of the circumstances of the urban poor in Bangladesh. All the 1979, 1990 and 1995 studies were directed by geographer-urbanist Nazrul Islam. The different dimensions of urban poverty were analyzed by a multidisciplinary team comprising senior geographers like Nazrul Islam, Rasie M Ahsan, Amanat Ullah Khan, Shanaj Huq-Hussain, Nurul Islam Nazem and others.

The poor in the city struggle against all odds, particularly against the state and the city authority and yet manage to survive, though at a very low level of human existence. The problem of housing provision in urban areas was very acutely felt in the 1970s and, along with problems of poverty, slums and squatters, housing received a tremendous response from researchers of many disciplines, especially architects, engineers, planners, demographers, statisticians and geographers (CUS 1980). The housing theme remained popular in research also into the 1980s and the 1990s (Islam and Ali 1991; Islam and Shafi 2008). Various aspects of the housing problem, including housing finance and affordability, and the role of different sectors, were addressed. Most of these studies relate to Dhaka. The public housing sector is fraught with institutional inadequacies, particularly financing problems. The formal private sector has been active since the 1980s and has now achieved some prospect in Dhaka; expanding to Chittagong and other major cities is still underdeveloped and small. It addresses mostly the needs of the upper-income groups (Islam et al. 1991). The informal private sector is the most effective one but unable to provide housing of reasonable standard. Research has shown possibilities in housing affordability of the low and middle-income groups. However, implementation of housing programmes have always been found to be difficult and a lack of political commitment has been identified as a major deterrent for the adoption of a national housing policy (Islam et al. 1991). Housing situations in some secondary cities have been studied by some geographers.

The problem of access to land for housing or urban development in general is even more basic than housing. The land question, however, received only limited attention in the 1970s but considerably more in the 1980s, 1990s and later decades. The rapid increase in land prices, unequal distribution of land ownership, insecure tenure of land, difficulty in land development, fringe land conversion and the problems of land administration were the issues highlighted in several studies (Islam 1990; Islam and Chowdhury 1992). The average price of land increased very rapidly in large cities, such as by 25 times in Dhaka between 1974 and 1989. The pattern of ownership of land has also been very skewed, particularly in Dhaka (Islam 1985–86) but also in small towns (Islam and Sadeque 1993). In Dhaka, the top 30 percent of the households have ownership or usage control over 80 percent of the residential land of the city (Islam 1985–86). Adoption of an urban land policy is a difficult exercise (Islam and Chowdhury 1992).

In general, the urban economy has not yet become a well-researched area for geographers in Bangladesh. The aspect of the urban economy which has received considerable emphasis is the informal sector. Urban industrial development and port studies received geographers attention in the 1950s and 1960s and also industrial clustered development very recently (ADB and CUS 2010). The traditional urban geographic approaches to the study of urban retail and other economic activities were carried out (e.g., Mollah 1973; Arephin and Ahmed 1990). The informal sector seems to be playing a dominant role in the urban economy of Bangladesh. More than 65 percent of Dhaka's employment was found to be in this sector in 1980 (Shankland Cox and Partnership 1981). Trading and transport absorbs the largest proportion of the informal sector employments (Islam and Khan 1988). Institutional aspects of urban centre also received research attention (CUS 1990; Islam N. et al. 1991). Urban or municipal administration suffer acutely from lack of coordination between various governmental and parastatal agencies. It also suffers from mismanagement and corruption (Islam et al. 2003).

Due to the heightened global awareness about environmental problems, the urban environment in Bangladesh is now becoming a major research theme (Khan, A.U. 1992). Some reasonably good work has been accomplished on the quality of urban water (Rob 1980; Rahman 1981; Shamsuddin and Alam 1988; Momtaz and Kabir 2014). Floods are life and death problems in Bangladesh. After the 1987 and 1988 deluges, the problem received enormous national and international attention. Flood also affects urban areas. The 1998 flood in Dhaka received the research attention of experts of several disciplines.

Geographers at CUS made several studies (Islam 2005; Dewan 2013). Riverbank erosion, another devastating natural hazard with impact on a number of urban centres, has been carefully researched by geographers at Jahangimagar University and the University of Manitoba, Canada (Elahi and Rogge 1991). Lack of funding and absence of necessary equipment and laboratory facilities stood in the way of conducting research on physical components of the environment. However, this gap is slowly being bridged by recent support of research in the field by the Ministry of Forest and Environment, Government of Bangladesh (Nazem 2015).

The quality of and access to urban infrastructural services like water, sanitation and energy have been investigated mostly through aided project consultancy (CUS 1985) but also as dissertations. Waste disposal also is a less studied area. A notable exception to this is research on garbage disposal (Ahsan et al. 1992). Research on infrastructure covered both Dhaka and several of the secondary cities. Both physical and planning aspects as well as socio-economic dimensions of the services have been explored.

Urban transport, like other infrastructures, have been studied by project consultants, but some academic research were also undertaken (Sharif 1994). The transport mode that dominates Bangladesh urban areas is the manually operated tri-cycle rickshaw and this has drawn a number of researchers to study its economic and social aspects. Light auto or para transport is also becoming a major mode in the cities of Bangladesh and its characteristics have been studied by some (Rashid 1981). Research on urban education has been limited (Banu 1981). Urban health has received fairly adequate research coverage. The major thrust of the attention, however, is on the medical aspects of the problem rather than on social, environmental and epidemiological aspects (Khan and Fariduddin 1989; Khan 1992; Khan and Huq-Hussain 2008). Intra-urban inequality in access to land, housing and other basic services has required some attention in recent years (Islam 1996b). The urban social structure has not been much of a research theme for geographers (Islam 1985–86). A beginning has also been made on the study of urban crimes in Bangladesh (Ahmad and Baqee 1988).

The status and role of women in the urban context came into particular focus in the early 1970s while the emphasis has grown in the following decades (for details, see Ahsan 1992). The structure of the urban labour force has in the last decade been greatly transformed by the entry of over 3 million women workers into garment industries, mostly in Dhaka. These women have been the object of some studies. Similarly, women in the informal sector have received research attention (Begum 1979; Huq-Hussain 1992). It seems that women perform mainly as a marginal segment of the labour force, constituting a reservoir of cheap labour which is largely tapped by various informal sector activities, whether as poorly paid wage workers or as unpaid family labour. Urban children, however, have received far less research attention.

The internal land use structure of cities has been a popular subject of research by geographers in Bangladesh since the late 1950s, but it was only very recently that an attempt was made to come up with a generalization on the structure. Based on a review of recent land use structures of 32 cities and towns of Bangladesh, Najia et al. (2008) have come up with two generalized models of land use: one dominated by the presence of a river and the other without a river and dominated by the railway line and roads. In either case, the presence of slums and squatter settlements feature prominently and so does the reality of agricultural land use within legal municipal limits. Most cities have a single city centre or CBD while mixed land uses are very common. The models compare well with land use patterns in many South Asian cities.

6. Concluding remarks

There are wide variations in the scope, methodology and quality of geographers' research on urbanization, there is a commonality in the selection of the study area. Most of the significant research works relate only to Dhaka, the reasons are many but, most obviously, limitation of funds. However, in recent years some attention has been given to cities like Rajshahi and Chittagong where universities have departments of geography. Geographic, physical planning, historical, sociological and more recently anthropological approaches are popular. Micro- rather than macro-level analysis characterize the scope of research. The majority of the studies were of an exploratory type. There is yet to be a major effort by geographers at building a theory of urbanization in a society deeply sunk in poverty but frantically trying to move up in the economic ladder.

References

Ahmad, N. 1957. 'The Urban Pattern in East Pakistan'. *Oriental Geographer*, 1(1): 33–41.
Ahmad, N. 1968. *An Economic Geography of East Pakistan*. London: Oxford University Press.
Ahmad, N. and M. A. Baqee. 1988. 'Urban Crimes in Bangladesh'. *Oriental Geographer*, 32(1 and 2): 65–74.

Ahmad, N. and A. K. M. H. Rahman. 1962. 'Development of Industries in Chittagong'. *Oriental Geographer*, 6(2): 149–158.
Ahsan, R. M., N. N. Haque and A. Haque. 1992. 'Waste Pickers and Recycling of Solid Waste in Dhaka City'. *Bangladesh Urban Studies*, 1(1 and 2).
Ahsan, R. M. and S. Sinowar. 1984. 'Mobility of Dhaka City Slum Dwellers: The Dynamics of Inner City Adjustment: A Case Study'. *Oriental Geographer*, 28(1 and 2): 14–29.
Arephin, S. and A. Ahmed. 1990. 'Retail Distribution in Rajshahi City: A Locational Analysis'. *The Journal of the Institute of Bangladesh Studies*, 13: 143–157.
Asian Development Bank (ADB) and Centre for Urban Studies (CUS). 2010. *City Cluster Economic Development: Bangladesh Case Study*. Manila: Asian Development Bank (ADB) and Centre for Urban Studies (CUS).
Bangladesh Bureau of Statistics (BBS). 2011. *Population and Housing Census 2011, National Volume-3: Urban Area Report*. Dhaka: Bangladesh Bureau of Statistics (BBS), Government of Bangladesh.
Bangladesh Bureau of Statistics (BBS). 2016. *Bangladesh National Accounts*. Dhaka: Bangladesh Bureau of Statistics (BBS), Government of Bangladesh.
Banu, S. 1981. *A Proposal for Integrated Metropolitan Secondary Education Authority for Dacca Metropolitan Area*. Unpublished Master's Thesis, Department of Geography, Dhaka University.
Begum, J. 1979. *Rural Urban Migration: A Survey on Poor Women in Two Communities in Dacca Metropolitan Area* (in Bengali). Research Report, National Council of Science and Technology, Government of Bangladesh.
Centre for Urban Studies (CUS). 1976. *Squatters in Bangladesh Cities*. Dhaka: Centre for Urban Studies, University of Dhaka.
Centre for Urban Studies (CUS). 1979. *The Urban Poor in Bangladesh: A Study on the Situation of the Urban Poor in Bangladesh With Special Reference to Mothers and Children*. Dhaka: Centre for Urban Studies, University of Dhaka (Sponsored by UNICEF).
Centre for Urban Studies (CUS). 1980. *Urban Housing and Shelter Process in Bangladesh: A Study in Dhaka, Chittagong, Khulna, Sylhet, Chandpur, Patuakhali and Thakurgaon*. Dhaka: Centre for Urban Studies, University of Dhaka.
Centre for Urban Studies (CUS). 1981. *The People of Dhaka*. Dhaka: Centre for Urban Studies, University of Dhaka.
Centre for Urban Studies (CUS). 1985. *Socio-Economic and Demographic Study of Six Districts Towns of Bangladesh*. CUS (6). Dhaka: Centre for Urban Studies, University of Dhaka.
Centre for Urban Studies (CUS). 1988. *Resettlement of 2600 Squatter Families at Mirpur*. Dhaka: Centre for Urban Studies, University of Dhaka.
Centre for Urban Studies (CUS). 1990. *The Urban Poor in Bangladesh*. Dhaka: Centre for Urban Studies.
Centre for Urban Studies (CUS). 1992. *Mapping and Survey of Slums and Squatters in Dhaka Metropolitan Area*. Dhaka: Centre for Urban Studies (with ICDDR, B).

Dewan, A. M. 2013. *Floods in a Megacity: Geospatial Techniques in Assessing Hazards, Risk and Vulnerability*. Dordrecht, Heidelberg, New York and London: Springer.
Elahi, K. M. 1972. 'Urbanization in Bangladesh in Bangladesh: A Geo-Demographic Study'. *Oriental Geographer*, 16(1): 1–15.
Elahi, K. M. (ed.). 1991. *Riverbank Erosion, Flood Hazard and Population Displacement in Bangladesh: An Overview*. Riverbank Erosion Impact Study. Savar, Dhaka: Jahangirnagar University.
Elahi, K. M. and J. R. Rogge, 1991. *The Riverbank Erosion Impact Study: Bangladesh*. Savar, Dhaka: REIS, Jahangirnagar University and Manitoba University, Riverbank Erosion Impact Study Project.
GoB, Planning Commission. 2011. *Sixth Five Year Plan*. Dhaka: Planning Commission of Bangladesh.
Hasnath, S. A. 1977. 'Consequences of Squatter Removal'. *Ekistics*, 44(263): 198–201.
Hossain, W. 1972. *Squatters in Dhaka City*. Unpublished Master's Thesis, Department of Geography, University of Dhaka.
Huq-Hussain, S. 1992. *Female Migrants Adaptation in Dhaka: A Case Study of the Process of Urban Socio-Economic Change*. Unpublished Ph.D. Thesis, SOAS, University of London.
Islam, N. 1975. 'Spacing of Urban Centres in Bangladesh'. *Oriental Geographer*, 19 and 20(1 and 2): 68–73.
Islam, N. 1985. 'Poor's Access to Residential Space in an Unfairly Structured City, Dhaka'. *Oriental Geographer*, 29 and 30.
Islam, N. 1986. 'Migrants in Dhaka Metropolitan Area'. In K. Husa, C. Vielhaber and H. Wohlschlagl (eds.), *Beitrage zur Bevolkerung. Vorschung*, pp. 285–298. Wien: Verlag Ferdinand Hurt.
Islam, N. 1990. *Human Settlements and Urban Development, Sector Report on National Conservation Strategy*. Dhaka: IUCN/ BRAC.
Islam, N. 1996a. 'City Study of Dhaka'. In J. Stubbs and G. Clarke (eds.), *Megacity Management in the Asian and Pacific Region*, Vol. 2, pp. 59–94. Manila: Asian Development Bank.
Islam, N. 1996b. *Dhaka: From City to Megacity*, Dhaka: Studies Programme, Department of Geography, University of Dhaka.
Islam, N. 2005. *Natural Hazards in Bangladesh: Studies in Perception, Impact and Copping Strategies*, pp. 19–28. Dhaka: Disaster Research Training and Management Centre, University of Dhaka.
Islam, N. and K. Ali. 1991. *Private Sector Initiatives in Urban Housing in Bangladesh*. Dhaka: Centre for Urban Studies.
Islam, N. and A. I. Chowdhury (eds.). 1992. *Urban Land Management in Bangladesh*. Dhaka: Ministry of Land, Government of Bangladesh.
Islam, N. and A. U. Khan. 1988. 'Increasing the Absorptive Capacity of Metropolitan Economies of Asia: A Case Study of Dhaka'. *Regional Development Dialogue*. 9(4): 107–133.
Islam, N. and H. Hossain. 1976. 'The Relationship of Urban Centres with their Rural Hinterlands'. In *Habitat: Bangladesh National Report*, pp. 67–76.

Vancouver: United Nations Conference on Human Settlements, May 31–June 14.
Islam, N. and A. Q. M. Mahbub. 1991. 'Growth of Slums in Dhaka City: A Spatio-Temporal Analysis'. In S. Ahmed (ed.), *Dhaka: Past, Present Future*, pp. 508–521. Dhaka: Asiatic Society of Bangladesh.
Islam, N. and A. Sadeque. 1993. 'Landownership in Small Towns of Bangladesh: Case of Sathkhira'. *Dhaka University Studies* (Bangla), 45: 75–90.
Islam, N. and S. A. Shafi. 2008. 'Use of Surface Water Bodies by Various Socio-Economic Groups in Dhaka City'. In T. Bertuzzo, G. Nest and S. A. Shafi (eds.), *Smooth and Striated, City and Water: Dhaka & Berlin*, pp. 59–71. Dhaka: Goethe-Institute Bangladesh.
Islam, N. et al. 1991. *Task Force on Social Implications of Urbanization*, Task Force Report (3). Dhaka: The University Press Limited.
Islam, N., M. M Khan, N. I. Nazem and M. H. Rahman. 2003. 'Reforming Governance in Dhaka, Bangladesh'. In P. L. McCarney and R. E. Stren (eds.), *Governance on the Ground*, pp. 194–199. Washington, DC: Woodrow Wilson Centre Press.
Khan, A. A. 1982. 'Rural-Urban Migration and Urbanization in Bangladesh'. *Geographical Review*, 72(4): 379–394.
Khan, A. U. 1992. 'Environment, Infrastructure and Health in Urban Areas in Bangladesh: A Research Review'. In N. Islam (ed.), *Urban Research in Bangladesh*, p. 155. Dhaka: Centre for Urban Studies.
Khan, A. U. and K. M. Fariduddin. 1989. 'Sector Report on Health and Nutrition'. In N. Islam (ed.), *The Urban Poor in Bangladesh*. Dhaka: Centre for Urban Studies.
Khan, A. U. and S. Huq-Hussain. 2008. *The Atlas of Urban Food Security Bangladesh*. Dhaka: GSRC and UN-WFP.
Khan, F. K. and M. Atiqullah. 1965. *Growth of Dacca City: 1908–1981*. Dacca: Social Science Research Project Publication, University of Dacca.
Khan, F. K. and N. Islam. 1964. 'High Class Residential Areas in Dacca City'. *Oriental Geographer*, 8(1): 1–40.
Khan, F. K. and M. Masood. 1962. 'Urban Structure of Comilla Town'. *Oriental Geographer*, 2(VI): 109–131.
Khan, S. A. 1972. *Transportation Routes in Bangladesh and their Impact in Trade Centres*. Unpublished Master's Thesis, Department of Urban & Regional Planning, Bangladesh University of Engineering & Technology, Dacca.
Mahbub, A. Q. M. 1986. *Population Mobility in Rural Bangladesh, the Circulation of Working People*. Unpublished Ph.D. Thesis in Geography, University of Canterbury, New Zealand.
Mahbub, A. Q. M. and H. Khatun. 1990. 'Urban Demographic and Social Structure'. In *The Urban Poor in Bangladesh*. Dhaka: Centre for Urban Studies.
Mahmood, A. Z. 1971. *Planning City Region: Case of Dacca*. Unpublished Master's Thesis, Department of Urban & Regional Planning, Bangladesh University of Engineering & Technology, Dacca.

Majid, R. 1970. 'The CBD of Dacca, Delimitation and the Internal Structure'. *Oriental Geographer*, 14(1): 44–63.
Mathur, O. P., N. Islam, D. Samanta and S. A. Shafi. 2013. *Sustainable Urbanization in Bangladesh: Delving into the Urbanization-Growth-Poverty Inter linkages*. Centre for Urban Studies (CUS), Dhaka.
Mollah, Md. Kh. I.1973. *Commercial Structure of Dacca City: With Special Reference to Retailing*. Unpublished Master's Thesis, Department of Geography, Dacca University.
Momtaz, S. and S. M. Z. Kabir. 2014. 'Environmental Problems and Governance'. In A. Dewan and R. Corner (eds.), *Geospatial Perspectives on Urbanization, Environment and Health*, pp. 283–309. Dordrecht, Heidelberg, New York and London: Springer Science & Business Media.
Muzzini, E. and G. Aparicio. 2013. *Bangladesh: The Path to Middle-Income Status From an Urban Perspective*. Washington, DC: The World Bank.
Najia, S. I., N. Islam and N. Rafiq. 2008. 'Urban Land Use Models: Case of Bangladesh'. *Bhugal O Poribesh Journal (Journal of Geography and Environment, in Bangla)*, 8: 37–57. Published in 2014.
Nazem, N. I. 2015. *Adaptation of Climate Change Induced and Environmentally Stressed Internally Displaced People (IDPs): A Study on the Dhaka Metropolitan Region*. Urban Studio, Department of Geography, University of Dhaka. Project Supported by Climate Change Trust Fund, Ministry of Forest and Environment, Government of Bangladesh (Ongoing study).
Rahman, H. 1981. *Evolution of Urban Community Development Programme in Bangladesh*. Dacca: University Grants Commission and Social Science Research Council.
Rashid, Md. A. 1981. *Intra-Urban Light Auto Transport: A Study in Dacca Metropolitan Area* (in Bengali). Unpublished Master's Thesis, Department of Geography, Dacca University.
Rob, M. A. 1980. *Water Supply and Consumption in Dhaka City*. M.Sc. Thesis, Geography Department, Dhaka University, Dhaka.
Saleheen, M. 1992. 'Characteristics of Rural Urban Migrants in Bangladesh'. In K. M. Elahi et al. (eds.), *Bangladesh: Geography, Environment and Development*. Dhaka: Bangladesh National Geographical Association.
Shamsuddin, S. D. and M. Alam. 1988. 'Industrialization and Urbanization Along Sitalakhya and Associated Pollution Problems: An Empirical Study'. *Oriental Geographer*, 32: 45–57.
Shankland Cox and Partnership. 1981. *Dacca Metropolitan Area Integrated Urban Development Project* (1–4). Dacca: Planning Commission, Government of Bangladesh, Asian Development Bank and United Nations Development Programme.
Sharif, A. H. M. R. 1994. 'Urban Transport Research in Bangladesh'. In N. Islam (ed.), *Urban Research in Bangladesh*, pp. 177–201. Dhaka: Centre for Urban Studies (CUS).

3 Dimensions of gender geography in Bangladesh

Rosie Majid Ahsan

1. Introduction

The word, 'gender', a relatively recent concept in social science, refers not to male and female but to masculine and feminine, i.e., to qualities or characteristics that society ascribes to each sex. Perceptions of gender are deeply rooted; they vary widely both within and between cultures and change over time. But in all cultures, gender determines power and resources for females and males. Gender refers to social inequalities between man and woman resulting from different variables across time, space and location (Oakley 1972). Gender Geography studies gender relations, description and the effects of gender inequality of space and place. Although both man and woman are thriving on this earth side-by-side, the study of women only emerged recently. This requires focusing on the history of geography itself in global context.

2. Background of gender geography

The first recorded use of the word 'geography' is found in the writing of Eratosthenes at Alexandria in the third century BCE. Here, 'geography' literally meant description of the earth, i.e., land and people of the inhabited world (Dickinson 1969). Location and interconnection of places were important. During the Age of Discovery, astronomy and mathematical and descriptive geography with latitudes and longitudes of places became important. By 1810, Carl Ritter introduced new geography in Germany. He claimed that the central principle of geography is the relation of all phenomena and forms of nature to the human species for identification of uniqueness of terrestrial areas in terms of land-human relations. He claimed that geography should be treated as a 'science of the earth'(Dickinson 1969).

In 1902, Paul Vilal de la Blache, the French geographer, dealt with the distribution of population and settlement and types and distribution of civilization (culture), and studied 'geographical environment and the human being'. According to him, the physical environment provided a range of possibilities which humans uses according to their needs and capacities.

Geography, in assessing the human-land relationship, thus studied human's activities in 'Economic geography', studied human's culture in 'Cultural Geography', but did not study the particular human being. In the 1950s, 'Population geography' emerged as a systematic branch of geography. In 1953, G. T. Trewartha argued that the study of population was the most vital element in geography and the one around which all the others were oriented and derived their meaning from – as such, it is essential to study the human for understanding spatial variation. Beajeu Garnier, J. (1956–57), Maximilian Sorre (1943–52), also contributed to the development of Population geography.

Population geography was the first branch to bring out women's impact on space through birth rate, death rate, population composition, etc. While studying the growth of people in the world, the quickening in population growth since the 1920s, resulting from improvement of technology (medicine and public health measures) and control in death rate, was observed especially in the underdeveloped countries of the East and South where birth rate remained high (Clark 1968). The question of population number vis-à-vis resources arose as the environment seemed to be exploited and damage and living conditions deteriorated. Problems of women's exploitation surfaced.

Social geography, emerging during the end of 1940s, gives genetic descriptions of social differences in relation to other factors. It deals with the differences in social groups and social problems of different areas (Jones 1975). But, it did not study women separately. It should, however, be mentioned that Hagerstrand did deal with women's paths while writing about the time-space path of the family in his essay, 'Definition of Migration', as an aspect of human behaviour.

3. Development programmes revealing women's importance: a global scenario

Evolution of science and technology in Europe over the past three centuries led to industrial pollution and devastating environmental situations through the process of chemical and radioactive contamination of soil, air and water leading to decreasing biodiversity and natural resources. This had a global impact, especially affecting poor countries.

To deal with the situation, various development projects were undertaken by Western agencies to helping and modernize the post-colonial societies. But they did not bring about the promised improvement in the living conditions of the people in the south and south-east. On the contrary, these projects led to the growth of poverty, gender inequality and the degradation of environment as they further reduced means of livelihood, especially for poor women. The growing ecological crisis in the world enhanced recognition of women's involvement in the struggle for survival. By the 1960s, international organizations such as the United Nations started declaring 'decades of development', which raised many questions.

The first decade of development in the 1960s aimed at rapid economic growth. It stressed industrialization of the South's post-independent countries (Braidotti et al. 1994) for economic progress which would be eventually profitable for the entire population. They perceived the role of women in reproduction (family planning) only as bearers and rearers of children. The second decade, in the 1970s realized that the trickledown effect did not take place, as industrialization affected the health and well-being of the families. Thus, it stressed development of people as a prerequisite for sustainable economic growth. Investment into human resources (loans) took place for equity and basic needs for improvement of health, education and social security. It had a brief note on integration of women in the developmental efforts but environmental destruction was overlooked. In the 1980s, the third decade attempted the removal of external imbalance of payment in the debtor countries, while social services were reduced, poorer people suffered and poor women suffered most.

The UN development decades saw women being oppressed in various societies of the world. During the 1960s and 1970s there emerged the 'Women's Liberation Movement in the West'. The decade of 1976–1985 was declared a 'Decade for Women' in an attempt to cope with the situation. The Women's Conferences in Mexico (1975), Copenhagen (1980) and Kenya (1985) all added to the struggle in the liberation of women. Debate on Women Environment and Development (WED) came up, which argues that women are positioned differently in relation to the use of environment. The ways that women are troubled by the deterioration of environment and the strategies they employ to combat the crisis are different. Publication of outstanding women such as Vandana Shiva saw women as environmental managers because of their nurturing capacity in regards to nature. Women could be treated as the source of a solution to the environmental crisis (Braidotti et al. 1994).

In the 1990s, the UN development decade was involved with environment and development. The United Nations Conference on Environment and Development, 'The Earth Summit', took place in Rio de Janeiro in 1992. In the process, an international women's meeting, namely, 'World Women Congress for Healthy Planet', involving NGOs, government agencies, business people and academia came up in Miami, 1991, producing a document, the *Women's Action Agenda 21*. This decade revealed the importance of women's roles in dealing with environmental problems. In 1995, the Women's Conference in Beijing made important contributions towards the emancipation of women. The decade of 1997–2006 was for the eradication of poverty, which took special care to promote gender equality and the advancement of women. Priorities include equal access to land, credit, job opportunities, equal legal rights and action to end violence against women.

4. Feminism: advocating equality of sexes

> Strength and self assertion lie within us all yet we are afraid, we do not use the resources given to us to lift our lives. Be true to yourself . . . dare to be different by courageously stepping forward to say, 'I am. I will be' then 'be'.
>
> (Asian Womanhood, Singapore)

The Women's Liberation Movement, advocating the previously mentioned philosophy, supports women's rights on the ground of equality of sexes. Feminism evolved to change these inequitable and unjust social relations.

Traditionally, women were viewed as different from men and historically women had less power, less control and less assets (Freeman 1994) as this was consistent with some natural order. Thus, Western science systematically excluded women from the actual activity of doing science and patriarchal scientists declared them to be unfit for usage of reason. According to them, women's domain was the household – the private sphere as opposed to the male domain of the public sphere. The scientific work of only white adult professional Western males was valued. As such, women were left out of development plans. Mainstream science is thus called Masculinist as it creates discriminatory hierarchies. The significant contributions of women in academic circles is of direct relevance in the struggle against mainstream science. The feminist perspective identifies similarities between men and women and declares that men are not superior to women and that women have the same potential for development. Academically, feminism has been part of politics for women, whether it is about control over reproductive rights or the construction of knowledge. In the early 1970s, Ester Boserup (1970), a liberal economist, highlights women's roles in the economic system. Her work had important contribution in Women in Development (WID) thinking. Her theme was an equal share for men and women. It demands legal equality of education and employment for women. WID believes in modernization and stresses Western values. However, the characterization of third world women was not always properly represented. Rounaq Jahan's attempt to make people aware of women's role in development – through a south and southeast Asian seminar in Bangladesh in 1977 – is noteworthy (Jahan and Papanek 1979). Gender and Development (GAD) came during the 1990s, which is concerned with increasing women's participation and benefits, looking for potential in development initiatives and empowering women. GAD focuses on social relations between men and women in the workplace and in other places. It views men as potential supporters of women and seeks structural reforms (Bishwanathan 1997).

5. Feminism in academia

Boserup's feminist ideas influenced various branches of social science subjects. Sociology was first in line, followed by linguistics, history, psychology, anthropology and political science. Feminist geographers came a little later. Feminist scholarship argues that mainstream science is customarily portrayed as universal, value-free and neutral in pursuit of truth. Feminists believe products of knowledge are best described as a social activity embedded in a certain culture and world view.

After the Women's Liberation Movement of the 1970s, in 1978, Eva Buff's Master's thesis for the first time treated women as a social group in geography in Germany (Baschlin 2002). Declaration of the UN decade for women (1975–85) was followed by the formal constitution of a 'Women and Geography' study group by the Institute of British Geographers in 1982 (Momsen and Townsend 1987). Formulation of the Commission on Gender by the International Geographical Union (IGU) took place in 1992, with Janet Momsen as its chair. A journal of feminist geography, *Gender Place and Culture*, has been published out of the UK since 1992, when it was founded by Liz Bondi and Mona Domash. Over the last four decades feminist geography has grown and challenged women's prescribed role within the household and social restrictions imposed on them. Geography, being involved with study of space in relation to humankind, has been interested in contributing to the development processes.

A number of waves of political approaches have been experienced by feminism in explaining women's lives. The first wave has been associated with 'social reform': abolition of slavery, demanding women's suffrage (right to vote), civil rights for black people, women's rights, gay/lesbian rights, bisexual rights and consciousness-raising. The second wave fought for equitable pay, sexual liberation, etc. The third wave is associated with difference and speaking from the margins, positioning the self in multiple oppressions (Moss 2002).

WID changed its approach through time, e.g., the 'Welfare Approach' started educating poor women about nutrition, family planning, etc. and women were treated as dependent on men; in the 'Equality Approach', women were agents and beneficiaries in all sectors but distribution of facilities was difficult to implement; the 'Anti-Poverty Approach' recognized that women need income generating measures for their basic needs. The 'Efficiency Approach' treated women as independent economic actors whose production potential was underutilized (Moss 2002). Women's social roles, however, were not given proper recognition in these approaches.

6. Feminist epistemology in geography

Ways of producing knowledge

An epistemology is a theory of knowledge with specific reference to the limits and validity of knowledge. An epistemology helps us to answer questions in finding truth/fact/reality in research.

Truth can be found through observation but many phenomena, such as 'democracy', cannot be seen. Knowledge is humanly constructed. As such, there are multiple perspectives coming from people of different standpoints and these perspectives are never value-free.

There are many philosophies in research that influence knowledge. Formulation of questions and ways of collecting data are all motivated by the researcher's belief. The gender of interviewers and respondents affect data. We try to minimize damaging prejudices and acknowledge other biases.

Feminist epistemology is a theory of know-how that investigates how taking gender into account affects what 'counts' as knowledge. The emphasis is to listen to women's voices. The ways the world is experienced by people is influenced by gender. Gender geography studies both oppression and privilege so that knowledge can be constructed differently. The multiple ways that gender affects society are always changing continuously affecting each other.

Feminist geographers started working with topics on spatial structures, as the relation between access to space and social power results in dominance or exclusion of social groups (Baschlin 2002). Feminist geography believes in the notion of context-specific and situation-sensitive knowledge rather than an all-encompassing truth. According to their study, mainstream science is not 'neutral' because it assumes that the norm for humanity is comprised of men only.

7. Gender geography: dealing with women's issues in spatial context

The term feminism has occupied an uneasy space within our societies in the East due to its association with 'lesbianism' which is not openly acceptable in our culture. Instead, 'gender' helps to explain a symbolic division of the sexes on the basis of systematic asymmetry between the sexes, which operates to the detriment of women (Braidotti et al. 1994), reflecting a logic of domination by man and exclusion of women from mainstream activities. Hence, the term 'gender geography'. In geography of gender, women's use of space, their relationship to place and the ways in which place and space constrain, and create opportunities for, women's life courses are studied (Katz and Monk 1993). Gender geography studies oppression of women in terms of patriarchy and capitalism, unequal access to existing political, social and economic institutions and resources. In a nutshell, geography of gender highlights the spatio-temporal variation of women-related issues, whether traditional or feminist (Ahsan 1998).

7.1 Methodologies in gender geography

Women's studies employ new procedures in understanding the problems, which would be more coherent than those offered by traditional academic disciplines. Here, the researcher is encouraged to share her own experiences and relate the problem to her own life, giving rise to a higher sensitivity in the approach. Masculine words such as 'mankind' are avoided because these words are status related.

The key concerns shaping geography of gender, as mentioned by (Moss 2002), are adopted from feminist geography:

(a) Analytical issues (can vary from issues which are pro-woman, anti-oppression or social justice). Adaptation of behavioural approach is suggested.

(b) Choice of data (data could be qualitative or quantitative as appropriate to the research questions).
(c) Scale of analysis (can vary from human body, people, home, institution, city or region. The spatial scale can vary from local, regional, national or international. However, local micro-scale studies are given greater attention).

8. Gender geography in Bangladesh

If we do not think of ourselves,
No one else will do,
Even if they do,
Hundred per cent benefit will not come to you.
So wake up sisters.

– Begum Rokeya

The previous quotation is that of Begum Rokeya Sakhawat Hossain (1880–1932), a Bengali Muslim woman litterateur, educationist and social reformer who played a pioneering role in awakening Muslim women in Bengal. She fought for female education in 1901, when girls were not given formal education and were mostly confined at home, with no recognized income earning activities and often married off at the age of nine. She learned to read and write Bengali from her brother and established the Sakhawat Memorial School for Muslim girls in Kolkata in 1911 with her husbands' money. Even women behind 'purdah' were encouraged to attend school as she provided transport, i.e., buses with special provision of 'purdah' with curtains on the bus windows. Since then, Muslim women of this region, who were especially deprived, have been advancing forward. Nojibunnessa of Rupganj thana, Dhaka, learned Bengali and Persian from her father as inspired by Begum Rokeya and set up a girl's school in her village in Masumabad around 1930.

Bangladesh is largely a Muslim country. Islam religion provides various rights to women, e.g., 'marriage rights' (women's consent is necessary in marriage), 'divorce rights', 'security for marriage' (alimony from husband – a husband should pay security money to the wife, which she can save or spend as according to her need), 'inheritance rights' (women have a right to inherit the property of her parents and her husband), 'remarriage rights' (a divorcee or a widow has a right to marry again) and 'prohibition of female infanticide'. Despite these protections for women, Islam has been misinterpreted and male preachers have abused religious regulations and arrested the physical, mental and psychological growth of women. Thus, despite the fact that, in Islam, a woman's position is high in the family, the practice was different in South Asia. Begum Rokeya created a momentum in the society for female education, employment and empowerment. A change in attitude among elite people resulted in some efforts towards empowering women in this region. Since the 1950s, women of this region have been given the constitutional right to vote and there was no discrimination in pay scale between male and female white-collar job holders. However, access to facilities for women was not equal in the society.

Bangladesh, with a population of about 160 million (2015) living in an area of 56,000 square miles, is basically an agricultural country where 48 percent of the people are female, of which 86 percent live in the poorer rural area. With its fertile land abundance of food supply and prosperous muslin trade with the West, Bangladesh was once a flourishing country. Its economic condition was first shattered due to the destruction of its muslin trade during the British period (1817). It has been continuously pressurized by frequent natural disasters, e.g., cyclone, flood, tidal surge and river erosion caused by its geographical location accompanied by political unrest. This led to problems of famine. Environmental decadence, the unfavourable alteration of our surrounding, co-existing with population explosion (as the death rate was controlled by Western technology) creates a fertile ground for poverty with severe impact on the most vulnerable section of the population – i.e., women, who became the poorest of the poor and are treated as commercial objects.

This situation is visible through beggars (mostly women), who were defeated in their battle against poverty and living on charity, and the trafficking of women and children emerging as a serious problem. Women and children from poorer sections of the population are being illegally sold outside the country to slavery or as sex workers. Within Bangladesh, also, many divorcees and deserted women and some very young girls from poor households are pushed towards prostitution.

With this background it is not surprising that women as an independent group drew our attention and became a subject of study in geography almost at the same time in Bangladesh as that in the West. *Researches on Migration and Fertility* (1984) by Nayar Sultana (Master's thesis), *Women's Role in Agriculture* (1987), resulting in a book, *Invisible Resource* through a joint effort by B. Wallace, R M. Ahsan, S. H. Hussain and E. Ahsan M. Phil, a thesis on *Fertility Pattern and Family Planning Practices* (1990) by Akhter Jahan were all undertaken, although not under the banner of gender geography. My training in population geography at the East West Centre Population Institute, Hawaii (1971–75), inspired me to generate research on female population as a research supervisor or as a direct researcher in Bangladesh. In Hawaii, I learned about Murray Chapman's ways of understanding the Melanesian people's behaviour, which had an indirect impact on my thoughts (although he is not a gender geographer). He developed an understanding of human behaviour in different cultural contexts. The poor women of Bangladesh also have a value system and behaviour different from that of upper-income women of the country and women from the West and require special understanding. As inspired by Hagerstrand, Chapman used a 'time-space graph' to illustrate spatial behaviour in the life cycle of individuals. Chapman's 'mobility register', with recorded events, was an effective information system, which required participant observation, generating greater sensitivity in analysis. Similar to Chapman, we used the 'work-time log' method in our first in-depth study of women (as advised by Ben Wallace, senior anthropologist of Southern Methodist University, USA) in Bangladesh, documenting 'minute by minute' and 'day to day' activities performed by rural women throughout the day, generating longitudinal data. Chapman's belief in micro studies was also an influencing factor. I was also convinced by his idea that geographers could contribute useful insights into developmental issues (Bedford 1999).

The first collective effort in generating research on women in Bangladesh geography was initiated in 1991 in the Department of Geography at the University of Dhaka, through an international seminar on 'Women and Environment' organized by the female faculty members (Nasreen Ahmad, Hafiza Khatun, Ammat-uz-Zohra Eusuf and myself), with full support of our senior geographers, Professor M. Aminul Islam and Professor K. B. Sajjadur Rasheed along with other male geographers. The seminar was co-sponsored by the Bangladesh Geographical Society (BGS) with Nazrul Islam as its President. It resulted in a special issue of *Oriental Geographer* (a journal of the society) on 'women' in 1992. I had the privilege of introducing the subject, 'Women and Environment in Geography', as the keynote speaker of the seminar.

Brigitte Silberstein, a French Geographer, and Leena Chatterjee from India were our initial contributors to gender geography. 30 geographers and other scientists and professionals from Bangladesh presented papers in the seminar on many relevant topics which, among others, included the following:

- Women's residential circumstance
- Natural hazards and vulnerability of women

- Women's health and spirulina – food security
- Gender disparity in education
- Violence on girl child
- Women and resources development
- Women and environmental pollution
- Women and social forestry

This created willingness among the geographers to consider gender as a new branch of study at the University of Dhaka.

In 1992, I started a course on 'Women and Development' at the M. Phil level, Department of Geography, University of Dhaka. Nasreen Ahmad helped in developing the theoretical background. The syllabus adopted for the course included the following:

- Emergence of women as a category in development: the spatial aspect
- Population dynamics: the gender aspect. Fertility, female mortality, female migration in urban and rural areas.
- Women and environment: as element and factor.
- Activities of women in the Developing World:

    (a) An evaluation of household activities
    (b) Agricultural activities and commercial activities
    (c) Industrial activities

- Activity space of women: rural and urban
- Women and basic needs: health, education, employment, housing in urban and rural areas
- Women and society: status, legal situation, patriarchal bargain, urban and rural area
- Women and institutions
- Management and interpersonal skills
- Gender Planning in the third world: meeting practical needs

Eventual activities included successful seminars on women one after another, which generated gender disaggregated data of this region. The Conference on Commercial Activities, Women and Ecology, was organized by the BGS, along with the IGU Commission on Commercial Activities in 1994. V. K. Shrivastava, Chairman of the IGU Commission on Commercial Activities, gave total support, along with his many female and male colleagues from India and Nepal. The seminar was graced by Professor Fazle Karim Khan and his colleague from Pakistan, Professor U. M. Malla from Nepal, Professor C. P. Singh, Professor

N. K. Dey of India and others. Professor K. B. Sajjadur Rasheed's keynote speech made a mark. The conference resulted in publication of a book, *Women, Work and Environment: Studies in Gender Geography*, edited by Rosie Majid Ahsan, Nasreen Ahmad, Ammat-uz-Zohra Eusuf and K. Nizamuddin in 1998. The book mainly published articles on women's commercial activities, along with the status of women and the degrading social environment.

In 1996, a book titled, *Female Migrants' Adaptation in Dhaka: A Case of the Process of Urban Socio-Economic Change* by Shahnaz Huq-Hussain added to the stream of publication (1996). It is a contribution from her PhD thesis. The millennium conference of BGS, on 'Disaster: Issues and Gender Perspective', held in December 2000, organized by Hafiza Khatun and me with the support of other members of the department added yet another book, *Disaster and the Silent Gender: Contemporary Studies in Geography* (2000). Hafiza Khatun and I did the editorial work. Since then a range of work has been done on many aspects of women and which is involved in gradually collecting data and contributing to the formulation of this branch of geography in Bangladesh.

A few important works published from the Department of Geography, University of Dhaka, have been mentioned here. Some papers (Malla 1998; Tooley et al. 2004) discussed why some women's groups differ from others in the way they utilize their resources. A number of articles studied women as protector and preserver of the environment (Nasreen 2004; Khatun 2005). Detailed analyses of some social problems faced by women in Bangladesh are dealt with in some other papers (Ahmad and Baquee 1991; Ahsan and Hossain 2004; Ahsan and Khatun, 2004). Women's health problems, which could be related to the attitudes and values of the society, along with problems of spatial environment, have also been studied (Rafique and Ahsan 1988; Khatun et al. 1994; Ahsan and Rezwana 2005). A large number of studies have been carried on women's activities and activity-space. These, among others, include women's productive activities as well as service-oriented activities (Baquee and Rasheed 1994; Eusuf 1994; Huq and Khatun 1994; Ahsan and Ahmad 1996; Ahsan et al. 2004). Studies have also been made about female garment workers and their residential problems (Islam and Haque 1991; Saleheen and Mia 1992). Over the last few years, geographers have paid attention to understanding various aspects of migration and their problems in particular in explaining the movement of different social groups within the female population (Ahsan 1997; Khatun 2000; Ahsan and Dey 2002).

Since 2004, a graduate level course on 'Gender Development and Environment' has been offered by Shahnaz Huq-Hussain at the Department of Geography, University of Dhaka, with the following syllabus:

- Concept and definition: gender, development and environment
- Gender-environment relationship (eco-feminism, gender and biodiversity, gender and forestry, gender and agriculture, gender and rural energy, gender and aquaculture)
- Gender and development: theoretical explanation
- Gender, production, reproduction and maintenance
- Gender, inequality and empowerment
- Gender and migration: type and dynamics
- Gender vulnerability (natural hazards: floods, cyclones and earthquakes; urban environmental problems: pollutions, waste disposal, health risks, trafficking)
- Gender and climate change: resilience and coping mechanisms
- Gender and globalization.

Interest in gender studies is also spreading to other universities in Bangladesh. At Jahangirnagar University a senior geographer Moudud Elahi showed his interest through his writings on gender relations in rural Bangladesh in Momsen and Kinnaird's book (1993).

Chittagong University introduced a course on 'Geography of Gender and Human Development' in 2005. It teaches the following:

- Geography of gender: definition, scope and linkages with related disciplines
- Gender disparity and gender roles in various societies; geographical differentials
- Women and men in production process, examples from UDC and DC
- Gender: resources and resource control
- Gender: economic activities and time use of women
- Gender and health: Family health and social issues
- Gender and global change, feminist, women liberation, women empowerment-geographic dimension from UDC and DC

The University of Dhaka set up a new department called Department of Women and Gender Studies in the year 2000, which offers courses such as 'Demography and Gender', 'Women Environment and Natural Resources Management', etc. I was a member of the syllabus committee. In the initial stage, Nasreen Ahmad, Hafiza Khatun and I taught in this department as part-time teachers.

In Bangladesh, gender geography does not strictly adhere to feminist principles in methodology, approach or philosophy. We believe in dealing with women's problems both through macro and micro studies with qualitative and/or quantitative data using our own experience if possible. Questionnaire survey, unstructured interviews, participant observation (formulating longitudinal data) and life history are used to collect the necessary data and facilitate sensitive analysis. We emphasize women's activities, both home based and public place based. Contributions from literature, culture, etc. are studied in relation to the environment. Gender problems, such as discrimination and exploitation in society, are dealt. Gender geography in Bangladesh differs from feminist geography in the participation of male geographers who recognize female problems and acknowledge the importance of women's studies and the advancement of women in the development of a country.

The primary purpose of developing gender geography in Bangladesh has been to demonstrate that women make a significant contribution to the economy and society of our country – in the rural areas, in agriculture and tea plantation and in the urban areas, presently (with the muslin industry in the past), in the garment industry, cottage industry, construction industry, food industry, education, health care, music, art, etc. and in miscellaneous services. Nevertheless, they are poor and suffer from oppression – but they have an ability to rebound from natural, political and social catastrophes. Success in developing a country, to a large extent, depends on well-planned national strategies. Gender geography in Bangladesh attempts to identify the vulnerable groups within society and carries out research in addressing their problems in the hope that it will make a significant contribution in shaping women's lives.

9. Putting gender geography in practice

On a practical level, to battle against gender inequalities and biases we must participate actively in the development plans. Awareness about neglect of women's basic needs and necessities in Bangladesh strongly advocates incorporating women into development processes. This is leading to an effort in upgrading women's status and liberating the society from inequalities and exploitation. Women play a major role in managing, caring and promoting natural resources, including raising children and developing the society, which is essential for sustainable development. The myth that rural women are not gainfully employed in Bangladesh has been broken. Social barriers against women's work outside the home broke down as the economic responsibilities of many households were given to women due to the mass killing of men in the liberation war of 1971. According to the Bangladesh Household Survey Capability Program Report, female-headed households are on the rise in Bangladesh and they are engaged in different forms of income earning activities.

Government plans and policies regarding women are becoming prominent, although it has been slow. In 1972, the Department of Women's Affairs was established under the Ministry of Children and Women Affairs. The first Five Year Plan (1973–78) had provisions for rehabilitating the war- affected abandoned women. The Second (1978–80) and Third Five Year Plan (1985–90) had provisions for women's employment and skill development programmes. From the mid-1980s, women were brought into the labour market. During the late 1970s, women labourers emerged as garment factory workers who contributed significantly to the economy of the country. Only in the Fourth Five Year Plan (1990–95) did the government aim at the integration of women into the development process. The National Policy for Women's Advancement (NPWA) was framed. WID in the Fourth Five Year Plan came during the Beijing Conference. It aimed at increasing women's participation in the public sector. The plan lays emphasis on raising productivity and income opportunity for women. Women's rights and privileges improved and education was encouraged.

The contribution of geographers to women and development started with the journalists' forum in the 1980swhere I was inspired to speak for gender issues and to advocate for the incorporation of women in the development agenda. A series of international seminars on women organized by the department of geography encouraged the generation of empirical data on women. Female geographers now contribute in different capacities and make valuable recommendations by using these data.

Addressing gender issues in designing and implementing development projects are thus given special attention where gender geographers are playing important roles. Academics are now encouraged by the university to be involved in public policy making and to participate in development planning. Women's problems are being recognized and prioritized and gender geographers' assistance is valued.

Development projects under different international agencies are in operation in Bangladesh, which are entirely devoted towards developing women. A few such names of projects, among others, are: (a) Vulnerable Group Development for Ultra Poor (VGDUP), (b) Integrated Food Assisted Development Programme (FADEP), (c)Food Security for Vulnerable Group Development (FSVGD) and(d) Food Security for Ultra Poor (FSUP).

These are all EU-implemented projects by different departments of the Government of Bangladesh and NGOs. The other similar projects funded by different donor countries, such as USA, UK, Japan, etc., have come to recognize gender development. The Rural Maintenance Programme (RMP) of the Ministry of Local Government is funded by the Canadian International Development Agency (CIDA) and implemented by the Cooperative for Assistance and Relief Everywhere (CARE). Canada gives employment opportunity to vulnerable, rural and poor women for year-round road maintenance in the rural areas.

The resettlement projects, such as the Jamuna, Padma, Bhairab river bridges and the Dhaka-Khulna Highway Action Plan, are all gender sensitive and demand contribution of gender specialists – this is where gender geographers' knowledge and services are required.

10. Concluding remarks

While gender inequality hinders economic development, it also contributes to the non-monetary aspects of poverty, i.e., lack of security, lack of opportunity and empowerment, lowered quality of life for both men and women. To plan a strategy of economic and social development without acknowledging women as a highly visible and powerful social and economic partner is to doom the plan to failure. This realization is ultimately inspiring policy makers and planners to make us women effective partners rather than treat us as a hindrance to the development of the country. Gender geography, which is creating this consciousness, is rising in importance as a subject in Bangladesh.

References

Ahmad, N. and A. Baquee. 1991. *Women and Violence, the Trend and Situation in Bangladesh*. Paper presented in the Seminar of the Bangladesh Geographical Society, Bangladesh.

Ahsan, R. M. 1997. 'Migration of Female Construction Labourers to Dhaka City, Bangladesh'. *International Journal of Population Geography*, 3(1): 49–61.

Ahsan, R. M. 1998. *Women Work and Environment: Studies in Gender Geography*. Dhaka: Bangladesh Geographical Society and IGU Commission on Commercial Activity.

Ahsan, R. M. and N. Ahmad. 1996. *Women in Commercial Activities: Beauty Parlour in Dhaka City*. Dhaka Folk Work and Place, Urban Studies Programme, University of Dhaka.

Ahsan, R. M. and S. Dey. 2002. 'Marriage Migration of Muslim and Hindu Women in Golachipa Pourashava in Patuakhali District: A Comparative Study'. *Oriental Geographer*, 46(1): 1–18.

Ahsan, R. M. and M. K. Hossain. 2004. 'Women and Child Trafficking in Bangladesh: A Social Disaster in the Backdrop of Natural Calamity'. In *Disaster and the Silent Gender; Contemporary Studies in Geography*. Dhaka: Bangladesh Geographical Society.

Ahsan, R. M. and H. Khatun. 2004. *Disaster and the Silent Gender: Contemporary Studies in Geography*. Dhaka: Bangladesh Geographical Society.

Ahsan, R. M. and N. Rezwana. 2005. 'Reproductive Health Problems of Poor Women in Tongi Slums in Two Villages of Gazipur District, A Geographical Analysis'. *Oriental Geographer*, 49(1): 1–20.

Baquee, A. and K. B. Rasheed. 1994. 'Women Traders in Periodic Market of Madhupur Forest Area Bangladesh'. *Women Work and Environment: Studies in Gender Geography*. Dhaka: Bangladesh Geographical Society and IGU Commission on Commercial Activities.

Baschlin, E. 2002. 'Feminist Geography in the German-speaking Academy: History of a Movement'. In *Feminist Geography in Practice*. Oxford: Blackwell Publishers.

Bedford, R. 1999. 'Mobility in Melanesia: Bigman Bilong Circulation'. *Asia Pacific Viewpoint*, 40(1): 3–17.

Bishwanathan, N. 1997. *The Women, Gender and Development Reader*. Dhaka: The University Press Limited.

Boserup, E. 1970. *Women's Role in Economic Development*. New York: St. Martin's Press.

Braidotti, R., E. Charkiewicz, S. Hausler and S. Wieringa (eds.). 1994. *Women, Environment and Sustainable Development: Towards a Theoretical Synthesis*. Santa Domingo: Zed Books.

Clark, J. I. 1968. *Population Geography*. London: Pergamon Press.

Dickinson, R. E. 1969. *The Makers of Modern Geography*. New York: Frederick A. Praeger Inc.

Eusuf, A. Z. 1994. 'Women's Involvement in Shrimp Industry: A Study of Selected Villages of South Western Coastal Area in Bangladesh'. In *Women Work and Environment: Studies in Gender Geography*. Dhaka: Bangladesh Geographical Society and IGU Commission on Commercial Activity.

Freeman, J. (ed.). 1994. *Woman: A Feminist Perspective*. California City: Mayfield Publishing Co.

Hagerstrand, T. 1957. 'On the Definition of Migration'. In E. Jones (ed.), *Readings in Social Geography*. London: Oxford University Press.

Huq, N. A. and H. Khatun. 1994. 'Cycle Rickshaw Repairing: Women With Toolkits'. *Gender Geography; A Reader: Bangladesh Perspective*.

Huq-Hussain, S. 1996. *Female Migrant's Adaptation in Dhaka: A Case of the Processes of Urban Socio-Economic Change*. Urban Studies Programme, No. 3, Dept. of Geography, University of Dhaka.

Islam, N. and S. O. Haque. 1991. 'Residential Circumstances of Low Income Single Working Women in Dacca City'. *Oriental Geographer*, 35(1 and 2): 48–63.

Jahan, A. 1990. *Fertility Pattern and Family Planning Practice*. Unpublished M. Phil Thesis, Department of Geography, University of Dhaka.

Jahan, R. and H. Papanek (eds.). 1979. *Women and Development: Perspective From South and South East Asia*. Dhaka: Bangladesh Institute of Law and International Affairs.

Jones, E. 1975. *Readings in Social Geography*. London: Oxford University Press.

Katz, C. and J. Monk (ed.). 1993. *Full Circle: Geographies of Women Over the Life Course*. London: Routledge.

Khatun, H. 2000. 'Role of Wife's Property in the Intra-Urban Residential Migration Pattern of Old Dhakaiya Households'. *Oriental Geographer*, 44, January.

Khatun, H. 2005. 'Flood Warning and Preparedness: Women Managing Flood in Bangladesh'. *Oriental Geographer*, 49(1): 61–73.

Khatun, H., L. Shamsuddin and S. Rouf. 1994. 'A Eclampsia: Result of Neglected Motherhood'. In *Women Work and Environment: Studies in Gender Geography*. Dhaka: Bangladesh Geographical Society and IGU Commission on Commercial Activities.

Malla, U. M. 1998. 'Women Ecology and Commercial Activity in South Asia'. In *Women Work and Environment: Studies in Gender Geography*. Dhaka: Bangladesh Geographical Society and IGU Commission on Commercial Activities.

Momsen, J. H. and V. Kinnaird (eds.). 1993. *Different Places Different Voices: Gender and Development in Africa, Asia and Latin America*. London: Routledge.

Momsen, J. H. and J. Townsend. 1987. *Geography Gender in the Third World*. London: Hutchinson.

Moss, P. (ed.). 2002. *Feminist Geography in Practice, Research and Method*. Oxford: Blackwell Publishers.

Nasreen, M. 2004. 'Coping Mechanism of Women in Bangladesh During Floods: Gender Perspective'. In *Disaster and the Silent Gender: Contemporary Studies in Geography*. Dhaka: Bangladesh Geographical Society.

Oakley, A. 1972. *Sex, Gender and Society*. London: Gower.

Rafique, N. and R. M. Ahsan. 1988. 'Age and Sex Differentials in Mortality in Rural and Urban Areas in Bangladesh: A Micro Study'. *Oriental Geographer*, 32(1 and 2): 99–114.

Saleheen, Mesbah-us and J. Mia. 1992. 'Women Workers in Export Oriented Garment Industries of Dhaka'. *Oriental Geographer*, 35(1 and 2): 85–93.

Sultana, N. 1984. *Migration and Fertility*. Master's Thesis (Dissertation), Department of Geography, University of Dhaka.

Tooley, M. and D. Dominey-Howes. 2004. 'Coastal and Sea Level Changes, Storms and Floods'. In *Disaster and the Silent Gender: Contemporary Studies in Geography*. Dhaka: Bangladesh Geographical Society.

Trewartha, G. T. 1953. 'The Case for Population Geography'. *Annals of Association of American Geographers*, 43(2): 71–97.

Wallace, B. J., R. M. Ahsan, S. Huq-Hussain and E. Ahsan. 1987. *The Invisible Resources: Women and Work in Rural Bangladesh*. Boulder and London: Westview Press.

4 Population geography in Bangladesh
In search of roots and approaches
K. Maudood Elahi

1. Introduction

This chapter starts with a brief evolutionary process of the subject 'population geography' in the West and then traces its trends of teaching and research in Bangladesh. During the mid-19th century, population was being studied initially as a matter of epidemiology. By the beginning of the 20th century, a separate discipline emerged as demography and then followed through spatial analysis – the core philosophical approach to geography – as spatial demography. In August 1854, cholera was a recurrent event in London, and Snow (1849) reviewed death records of residents who died from cholera and interviewed household members. He determined that most of the people having died from cholera lived near and drank water from a pump in the locality. Snow presented his findings to community leaders and the pump handle was removed in the following month, preventing additional cholera deaths. Often Snow is heralded as a legendary figure in epidemiology because he provided one of the earliest examples of using epidemiologic methods to study factors affecting the health and disease of populations.

Other early scientists used similar methods of mapping spatial distribution to examine demographic phenomena (Mayhew 1861; Booth 1902). At a much later date, a similar study of mapping poverty in the slums of Chicago was done by Florence Kelley using cartograms to represent density of family members in housing areas. As these studies indicate, maps have enormous analytical potential. Traditionally, maps have served as a valuable role in understanding demographic characteristics and events. Maps represent the geographic epistemology – thinking spatially and thereby adding characteristics to population analysis to date.

2. Evolution of population geography

2.1 Conceptual and theoretical approaches

Although geographers have long been interested in population with a focus on spatial and temporal contents, population geography as a discipline has a short story and in Bangladesh a much shorter one. In this context, Trewartha (1953), a noted American climatologist, is thought to have proposed the idea of a separate field with a major focus on 'human population'. In his presidential address, titled 'Á case for population geography', at the annual meeting of the Association of American Geographers (AAG) in Cleveland, Ohio in 1953, Trewartha argued that 'the study of population, long neglected by the discipline, deserved a more prominent position in geography's agenda' (Trewartha, 1953). Trewartha also proposed a very comprehensive outline of the content of this subject, which many geographers both in the West and East have retained for several decades. In 1969, he qualified his earlier definition of population geography as concerned with the understanding of the regional differences in the earth's covering people (Trewartha 1969). Nevertheless, the subject has been a matter of debate ever since Trewartha formally raised the issue in 1953, as is the case with the definition of the sub-discipline.

After an embryonic decade or so, two books on population geography appeared in 1966 – one by Zelinsky (1966) and the other by Beaujeu-Garnier (1966). This was the time geography as a discipline was redefined in various ways, such as 'the study of pattern and process associated with the earth'; 'the study of relationships between humans and their environment by emphasizing a spatial and environmental perspective at a variety of scales'; and 'a spatial discipline . . . that seeks to understand pattern of earth and the processes that created them' (www2.una.edu/geography/statedepted/definitions.html). It may be noted that in all the definitions, the contexts of time, space and environment were emphasized in relation to human existence. The present author likes to define geography as 'the discipline that studies the relationship between humans and their environment through space and time'. With the reference to time and space, the definition of population geography was also redefined, and as Zelinsky (1966) puts:

> Population geography can be defined accurately as the science that deals with the ways in which the geographic character of places is formed by, and in turn reacts upon, a set of population

phenomena that vary within through both space and time as they follow their own behavioral laws, interacting one with another and with numerous non-demographic phenomena.

Although population geography is heavily dependent on demographic data, it starkly contrasts with the definition of demography which is the statistical and mathematical study of elements of population.

Clarke (1972) suggested that population geography is mainly concerned with demonstrating how spatial variation in population and its various attributes like composition, migration and growth are related to the spatial variation in the nature of places. Zelinsky (1966), a contemporary of Clarke, takes a similar view regarding the definition of population geography. Similarly, Pryor (1984) held that population geography deals with the analysis and explanation of the interrelationship between population phenomena and the geographical character of places as they both vary over space and time.

Although Trewartha and Zelinsky made an impact on the discipline of geography as far as its conceptual and methodological status, Hauser and Duncan (1959), as demographers, in their classical book earlier established a concrete linkage between the subjects, making population geography an effective interdisciplinary sub-field.

2.2 Methodological aspects

In addition to the growing recognition of the significance of population elements in geography, some other developments that were taking place in different parts of the world, particularly during the 1970s and 1980s, helped a great deal in the emergence of 'sub-fields' of population geography. This new development led to a number of flourishing research areas in the following several decades. As Clarke (1984) has suggested, the growing availability of population statistics by a number of national and international agencies has played a crucial role in the emergence of population geography. The publishing of demographic data of the world by UN agencies allowed research at spatial and temporal levels by population geographers in various countries, mostly at macro-scale.

The need for a more detailed account of other demographic data resulted in a switch over from macro- to micro-level studies, which, in turn, facilitated population mapping and analysis (Kosinski and Webb 1976,1977; Clark 1984). The activities of the Population Commission of the International Geographical Union (IGU) allowed the expansion of population studies by the geographers during the 1970s until the end of the last century. The activities of the IGU Population Commission received added impetus from L. A. Kosiniski (Alberta University, Canada) and John I. Clarke (Durham University, UK) during which the present author from Bangladesh acted as a regular member of the Commission from1976 to 1980 and 1980 to 1984 (see Figure 4.1).

It was the period when the Commission organized about six international symposia (Minsk, Karachi, Katmandu, Tokyo, Edmonton and Zaria) bringing together population geographers from around the world and facilitating academic and research interactions resulting in more than a dozen publications highlighting the work of population geographers. During this period, the Commission on Population Geography maintained a close liaison with the Population Council, the International Union for the Scientific Study of Population (IUSSP), the International Labor Organization, the United Nations Population Fund (UNFPA), etc.

Figure 4.1 The author with some of the leading population and human geographers of the post-1950s at the Symposium of the IGU Commission on Population Geography held in Zaria, Nigeria, 1978 (left to right: K. Maudood Elahi (the author), J. I. Clarke (UK), A. Mondjanngni (Cameroon), R. J. Harrison-Church (UK) and L. A. Kosinski (Canada))

Source: Author

3. Population geography in Bangladesh

3.1 Major theoretical framework

Much of the late-19th century and early 20th century population was studied as a part of Economic/Commercial Geography in the West. Subsequently, it was dealt with as a part of human geography. For several decades, it was defined as 'a division of human geography that focuses on how the migration, distribution and the growth of population is affected by the nature of a place'. During the initial years of the 20th century, as in the West, most South Asian universities used to teach key issues of population within the framework of economic geography or human geography. During this period the Aligarh Muslim University (India) took the lead in this regard. At about the same time, issues of human settlements also gained interest in geographic studies in some universities with a central focus on population, and the Banaras Hindu University (India) took the lead in the process. With the establishment of Dhaka University in then East Bengal in 1919, teachers, mostly graduated from the Aligarh Muslim University, carried on the tradition of this university to teach geography while joining in the university too. They started teaching population and settlement geography until themid-20th century. The only pioneering personality in developing the sub-field of population and settlement was Fazle Karim Khan (Dhaka University) in this period.

The late-1960s and 1970s also saw the increasing use of quantification aided by access to computers that helped geographers handle large data. The main concerns of population geography revolved around the following aspects of human population:

- Size and distribution (rural and urban at both micro- and macro-scales)
- Population dynamics (fertility, mortality and migration)
- Population composition (ascribed and non-ascribed characteristics)

A number of doctoral research works on these issues resulted for the first time in Bangladesh. The key researchers from Bangladesh were: K. Maudood Elahi (Durham University, 1971), Rosie Majid Ahsan (University of Hawaii, 1974) and Alak Paul (Durham University, 2009). In particular, Elahi, for the first time, made use of the Multivariate Model in his doctoral research to analyze various population characteristics. At about the same time, studies on migration issues created interest among population geographers around the world, including

Bangladesh. Besides, there were numerous postgraduate/MS studies done at home and abroad.

It was only after 1971, as Bangladesh emerged as an independent country, that much of the existing curricula of geography underwent drastic revision, incorporating newer subject matters and changing the name of the discipline to 'Geography and Environment', in keeping pace with the change in the existing curricula. In this connection, the Department of Geography at Jahangirnagar University took a major lead by introducing population geography as a separate course at both the BA and MA levels along with the introduction of quantitative geography, settlement geography and housing, Geographical Information System (GIS) and Remote Sensing, etc., in order to further strengthen the methodological basis of this sub-field. In this university, K. Maudood Elahi, supported by AFM Kamaluddin and Mesbah-us-Saleheen, played a key role in introducing population geography along with the previously mentioned courses. Similarly, the Rajshahi University followed Jahangirnagar University at the initiative of Sirajul Arefin, Jafar Reza Khan, Abu Taha and Shamsul Alam, who played a key role in developing studies of population geography and other sub-disciplines. The Dhaka University followed this trend at a somewhat later stage at the initiative of M. Aminul Islam, Nazrul Islam, K. B. Sajjadur Rashid and others.

In a much later period, the Chittagong University established the Department of Geography and Environment in 1998. This Department basically followed the existing course content of population geography practiced in Jahangirnagar University as did the Begum Rokeya University at Rangpur in 2009 at the behest of K. Maudood Elahi (Stamford University Bangladesh) and Zahid Hasan (Rajshahi University).

3.2 Contemporary research interests

Since the mid-20th century there was also a growing concern among the geographers (as well as national governments and people at large) about population expansion and its effect on economic development. Bangladesh and many South Asian countries have been experiencing redistribution of population within their boundaries (Kosinski and Elahi 1985). Further, the impending impacts of climate change and sea level rise as a result of global warming created newer avenues of research by geographers as the consequences tend to encompass internal displacement of population, cross-boundary population movement, refugee problem, newer diseases regime and a number of regional political and security tension due to slowly emerging climate/environmental refugees (Elahi et al. 1999; Elahi 2007; Elahi and Sultana 2010; Elahi 2016). These are considered as the newer frontiers of research focus by the population geographers of Bangladesh for the rest of this century.

As in the West, the courses on population geography influenced other areas of geography during this period. Some examples are: medical geography, geography of health, settlement and housing geography, migration geography, human ecology, population and sustainable development, etc.

To summarize, the pioneer exponents of population geography held that population is the point of reference from which all other elements may be observed in order to seek meaning and significance. Later, up to the 1970s, population geography was characterized by the quantification of population elements to understand spatial structure over time. This was the most prosperous period for population geographers of Bangladesh as they began their higher training or studies from universities abroad. The third stage lasted from the 1970s to the 1990s, during which the Bangladeshi population geographers maintained close liaison with their counterparts in UK, US, Canada and India. As a result, a number of applied and empirical research projects ensued, particularly at Jahangirnagar University (Rogge and Elahi 1990). Most of the departments of geography and environment in other universities started focusing on environmental linkages in their curricula of population geography. The fourth stage lasted thereafter until the present time, when newer universities opened departments of geography and environment, and development and application of GIS and computer-aided data analysis made practical inroads into research methodology by some geographers. A number of trained individuals for research and development joined as teachers in Bangladesh Universities at this stage. In this period, spatial analysis and quantitative methods were reaffirmed by the geographers; due to this orientation, the methodology of some pure and social demographic issues underwent restructuring, too, and a number of interdisciplinary and applied/empirical researches followed mainly at Jahangirnagar University.

4. Population geography-development linkage

CHALLENGES AND FUTURE DIRECTIONS

Despite the importance of population geography, the contribution of geographers have largely been limited for two reasons: (a) a smaller number of population geographers working on relatively selected issues and (b) an increasing number of researchers from population-related or allied fields of social sciences (like sociology, gender studies, economics, social anthropology/anthropogeography, applied demography, etc.). It has been seen that present-day geographers are also lagging behind the latter due to non-innovative methodological interventions in their research. Therefore, the geographers are likely to fail in meeting the planning mandates of the Sustainable Development Goals (SDG) of the country in recent decades unless they address the gap in a concrete way.

The main purpose of these planning mandates is to improve the standard of living or welfare of the people on the basis of the available population-resources scenario. But this population-resources relationship is a dynamic one, both in time and space. It is in this context that there are some unaddressed issues as research interventions in population studies as well as some potential areas of resource utilization for geographers. Some of the major interventions may be noted here. Currently, the geographers dealing with population greatly suffer from methodological deficiency in the sense that: (a) some tend to avoid a statistical basis of sampling frame in their research design, (b) most have failed to make good use of computer-aided statistical tools (for example, Statistical Package for the Social Sciences (SPSS), STATA, etc.) (c) there is an inappropriate use of GIS techniques (thinking that they are only mapping tools), and (d) there is a failure to integrate their research outputs with development planning rhetoric for the previous three reasons (a to c).

5. Concluding remarks

It may be noted that delineating the precise field of the sub-discipline, 'population geography', in Bangladesh universities has been a major problem since its beginning. This might be because of the fact that population geographers are too limited in number and some have not been able to establish themselves in the research niche of population studies in the country in a significant way. On the other hand, there have not been sufficient research activities on applied issues of population by the geographers, even though such opportunities are plentiful with the emerging problems of climate change and development in Bangladesh.

For Bangladesh universities, population geography is now no longer a compromised field of spatial and temporal applications of dynamics of population (i.e., fertility, mortality and migration) only. Contemporary population geography is theoretically sophisticated, integrating spatial analysis, GIS and geo-referenced database and quantitative analysis with reference to planning and development needs. Future progress in the field of population geography needs to be developed with more applied and empirical research on planning and societal development issues. Some such research areas may be related to population growth and development; population displacement, migration and refugees in response to natural disasters; climate change-health-environment nexus, especially extreme climatic events and population displacement, migration and environmental refugees; vulnerability studies and food security; development and energy security; sustainable development and poverty reduction; and transnational human security issues. These and similar other related issues will very rightly form part of the overall concern of population geographers in the present time.

References

Beaujeu-Garnier, J. 1966. *Geography of Population*. London: Longman.
Booth, C. 1902. *An Early Account of Mapping and Demographic Analysis*. London: MacMillan & Co.
Clark, W. A. V. 1984. *Population Geography at Micro-Scale: Residential Mobility and Public Policy*. Oxford: Pergamum Press.
Clarke, J. I. 1972. *Population Geography*. Oxford: Pergamon.
Clarke, J. I. 1984. *Geography and Population: Approaches and Applications*. Oxford: Pergamon.
Elahi, K. M. 2007. *Migration and Environment: The Effect of Extreme Environmental Events* (Keynote paper for Session III: The Effect of Extreme Environmental Events on Migration). Expert Seminar on Migration and the Environment, International Organization on Migration (IOM), Bangkok.
Elahi, K. M. 2016. 'Climate Change and Health Impacts in Bangladesh'. In R. Akhtar (ed.), *Climate Change and Human Health Scenario in South and Southeast Asia*, pp. 207–219. Cham: Springer.
Elahi, K. M., K. S. Ahmed and M. Mafizuddin (eds.). 1999. *Riverbank Erosion, Flood and Population Displacement in Bangladesh*. Dhaka: REIS-JU-UM.
Elahi, K. M. and S. Sultana. 2010. 'Resurgence of Malaria in Bangladesh'. In R. Akhtar et al. (eds.),*Malaria in South Asia*, pp. 107–122. New York: Springer.
Hauser, P. and O. D. Duncan (eds.). 1959. *The Study of Population: An Inventory and Appraisal*. Chicago: The University of Chicago Press.
Kosinski, L. A. and K. M. Elahi. 1985. *Development and Population Redistribution in South Asia*. Dordrecht, Boston and Lancaster: D. Reidel Publications. (Reprinted: Jaipur, New Delhi: Rawat Publications, 1991).
Kosinski, L. A. and J. W. Webb (eds.). 1976. *Population at Micro-scale*. Hamilton: New Zealand Geographical Society.

Kosinski, L. A. and J. W. Webb (eds.). 1977. *Population and Scale: Macro-Population*. Edmonton: IGU Commission on Population Geography.

Mayhew, H. 1861. *London Labor and the London Poor: A Cyclopedia of the Condition and Earnings of Those Who Do Not Work, Those That Cannot Work, and Those That Will Not Work*. London: Griffin, Bohn, and Company.

Pryor, R. J. 1984. *Methodological Problems in Population Geography*. Oxford: Pergamon.

Rogge J. and K. M. Elahi. 1990. *Riverbank Erosion, Flood and Population Displacement in Bangladesh: A Report on the Riverbank Erosion Impact Study (REIS)*. Dhaka: REIS-JU-UM.

Snow, J. 1849. *On the Mode of Communication of Cholera*. London: J. Churchill.

Trewartha, G. T. 1953. 'A Case for Population Geography'. *Annals of the Association of American Geographers*, 43(2): 71–97.

Trewartha, G. T. 1969. *Geography of Population: World Patterns*. New York: John Wiley.

Zelinsky, W. 1966. *A Prologue to Population Geography*. Princeton: Prentice-Hall.

5 Geographies of HIV/AIDS in Bangladesh

Global perspectives, local contexts and research gaps

Alak Paul

1. Introduction

Different disciplines, including geography, offer varying perspectives on health and society. Over time, as for many other disciplines, fashions and emphases change in geography (Kearns and Moon 2002). These changes have long been recognized as the development of knowledge and scientific thought. Thus, a quickening process of metamorphosis has shifted 'medical geography' from a minority concern and a 'confusing sub-variety' of human geography to a confident, recognized and distinct 'geography of health' (Rosenberg 1998). Many recent works in the branch of health geography have suggested that a shift from medical geography to the geography, or geographies, of health is more than a change in title. It also represents an epistemological shift that questions the grounds upon which medical geographical knowledge is based (Gesler and Kearns 2002). Distinctions are often made between medical and health geography, with medical geography dominated by biomedical models favouring intensive quantitative methods, whereas health geography relies more on socio-ecological models and often employs more extensive qualitative approaches (Gesler 2006).

Geographers have worked on infectious diseases such as cholera, malaria, influenza, measles and hepatitis for a long time (for example, May 1958; Stamp 1964; Pyle 1969) and are now making contributions to the geography of HIV/AIDS in the contemporary period. From the early years of the HIV/AIDS epidemic, it was apparent that international travel – for business and for pleasure – played a necessary, though insufficient, role in the geographical and social diffusion of HIV infection (Gould 1993). Geographers have given their attention to the virus itself, developing diffusion models or recounting origins theories with no reference to the spatial-political implications of living with AIDS (Kearns 1996). They have also developed predictive models of future diffusion patterns and epidemiological changes across the regions (Takahashi 1997). By contrast, some geographers have looked at the social context in which HIV gets transmitted and through which particular individuals become vulnerable. Both of these focuses (quantitative and qualitative) are discussed in detail in what follows along with a discussion of a research gap in the geographies of HIV/AIDS in Bangladesh.

2. Quantitative focus

Analysis of the geographic distribution and migration of HIV is an established field of study where mapping is the main element to show the origin of the virus and its diffusion over space. In 1989, Shannon and Pyle published a paper where they provided an overview of AIDS from the medical geographical point of view. They explored questions pertaining to the origin, aetiology and spatial diffusion of the disease by using available hypotheses, theories and data. The detailed work of Wood (1988) and Smallman-Raynor and Cliff (1990) are also in the tradition of quantitative medical geography. They looked at hypotheses of the origins of AIDS and its spread. Gould (1989) discussed the geographic dimension of the AIDS epidemic on the basis of Wood's (1988) article and agreed with his spatial diffusion pattern of AIDS. Smallman-Raynor and Cliff (1990) described the Gallo (1987) model of major global spread of HIV-1, which hypothesizes major diffusion corridors from the areas of endemic infection in west and central Africa to the rest of the globe between the late 1970s and mid-1980s. Cliff and Smallman-Raynor (1992) explored the geographical variation in the incidence of AIDS in Uganda and they considered three hypotheses in the context of spatial variation of the virus. Gould and Wallace (1994) have researched the spread of HIV by mapping the cumulative incidence of AIDS cases across the USA and they have also analyzed the number of reported cases in the New York region. Cliff and Haggett (1988) included AIDS in their book in relation to the complexity of its spread due to a long incubation period, its current lack of cure, its high virulence and because it results from one of the most basic urges of the human race. Dutt et al. (1987) performed a spatial analysis of HIV/AIDS in the United States by region, by patient, gender and age groups of the victims, while Paul (1994) examined the spatial pattern of AIDS distribution in Asia. To explore the 'sociogeographic' network of HIV infection, Wallace (1991) suggested a useful model for the initial spread of HIV in a social or a geographic network where individuals are treated as nodes and relationships – as the connecting links between individuals. All of these previously mentioned works have emphasized mapping as a means of exploring the origin of the HIV virus and its diffusion.

Many studies of migration and HIV/AIDS have shown considerable mobility among people with HIV and have identified key mobility (from urban-rural or rural-urban) factors, such as the need for social services, family support, changes in lifestyle, avoidance of high risk behaviours and the need for access to better health care (Lam and Liu 1996). Recently, a number of studies have been made to understand the impetus and reasons for migration among persons with HIV (Wood et al. 2000). In the context of sub-Saharan Africa, Oppong (1998) and Mayer (2005) have reviewed the spatial perspective of HIV/AIDS. Geographers have also drawn attention to the impact of HIV on the individual behaviour and conditions of life from an ecological perspective, while Aase and Agyei-Mensah (2005) highlighted the importance of place and gender. Kalipeni et al. (2004) shifted the focus of work on HIV/AIDS in Africa from purely epidemiological approaches towards a social science perspective on the spread of the disease. Teye (2005) discusses a risk theory that argues that the individual is not a free agent when making choices due to constraints on a societal level. Piper and Yeoh (2005) described the HIV epidemic in Asia as more complex than that in Africa, involving a multiplicity of transmission modes. Moran (2005) worked on 'new wave' of HIV epidemics in Russia with transitional issues of risk and vulnerability. However, although much empirical research shows a connection between migration and HIV spread, few theoretical works have tried to connect the social and behavioural mechanisms (for example, Lindquist 2005; Marten 2005). In shaping the recent geography of new HIV infections, Ferguson and Morris (2007) highlighted the roles of 'vulnerable places'; Kalipeni and Zulu (2008) offered an alternative method to predict future HIV prevalence; Oppong and Harold (2009) depicted an equal role of HIV vulnerability by the people and place; Feldacker et al. (2010) discussed spatial associations; Vearey et al.(2010) explored the social determinants of urban health; Ezekiel et al. (2010)examined the interrelationships of social, spatial and temporal issues; Magadi and Desta (2011) explained the patterns and factors of HIV seropositivity; Madise et al. (2012) highlighted intra-urban differences; Yao et al. (2014) examined the impact of a rapid expansion of HIV-related services on access to and utilization of HIV testing; Carrel et al. (2016) considered urban proximity as a transnational phenomenon; and Ransome et al. (2016) examined late HIV diagnosis rates. Here, all the authors used different methods or tools like Geographically Weighted Regression (GWR), GIS-based methods, WebDMAP, kernel density method, multilevel regression analysis, etc. to quantify the HIV prevalence or progression in their respective study place or region. All of the quantitative-based works have determined that HIV/AIDS has tended to cluster in certain areas and infection, diffusion and overlap can be expected between different population subgroups, even those who have not traditionally been at risk. This migration analysis is one of the best ways to anticipate how the epidemic will evolve and hence to direct the necessary steps to try and avoid further spread.

3. Qualitative focus

Although a large body of literature currently exists on HIV/AIDS in general, geographical studies focusing upon the experience of people with HIV/AIDS are limited. In the 1990s, geographers began to make contributions to an understanding of the AIDS epidemic in the form of mapping to illustrate the origin of the HIV and its diffusion at global, national and local levels. These contributions of spatial diffusion of the virus in the geography of AIDS are valuable and important but provide only a 'single and partial geography of AIDS' (Wilton 1996). Relatively little attention has been paid to the socio-cultural, economic and political dimensions of the disease or to the ways in which individual risk behaviours are related to non-spatial processes in particular geographical settings (Asthana 1998).

In this regard, alternative geographies have started to document the appearance of HIV, its progression to AIDS, and its manifold complications, which undoubtedly affect the geography of HIV-positive people's daily lives. One promising response to this research gap is Wilton's (1996) ethnographic study of the daily life experiences of Los Angeles men with symptomatic HIV/AIDS. This work disaggregates the social, physical and psychological dimensions that help him to show how place matters in the lives of these men. In his subsequent paper, Wilton gives a framework of qualitative research with a critique (1999). Brown (1995) also carried out ethnographic work on gay men's experience with HIV/AIDS and argued that geographers' preoccupation with the spatial diffusion of the virus threatens to reduce gay men's bodies to biological hosts, a subject of interest only to the extent that they are able to transport HIV across space. In another piece of empirical work that included a distinctively ethnographic approach to health, Brown (1997) looked at the local politics of AIDS in Vancouver, Canada and tracked the location of political engagement with the AIDS crisis. Sabatier (1996) conducted interviews with migrant women from southern Africa, examining their vulnerability to HIV infection together with the mechanisms they adopt for coping with the infection. Asthana and Oostvogels (1996) carried out ethnographic work with commercial sex workers regarding their problems and prospects for participation in HIV prevention. Craddock (2000) focused upon the role of social identities and cultural framings of diseases particularly calculations of disease risk while Kesby (2000) explored gender relations for HIV patients in rural Zimbabwe. Browne and Barrett (2001) have raised the question of how states define and represent norms of sexual morality and the AIDS pandemic has forced states to redraw the 'moral boundaries' of consideration and acceptability for public consumption.

Geographers have begun to explore the nature and extent of the impact of HIV/AIDS on health services and on diverse aspects of the daily lives of people living with HIV and AIDS (for example, Takahashi and Smutny 2001; Del Casino 2001; Young and Ansell 2003; Valentine 2007; Brown 2012). Some studies have explored the social construction of HIV/AIDS (for example, Alonzo and Reynolds 1995) where shifting stigma, identity change and social acceptance are discussed. Discourses around health care, sexuality, gender and migration related to the epidemiology of HIV/AIDS and local processes involving individual and community response are much argued about by geographers, anthropologists and others. These concepts are also related to class, gender, sexuality, culture and politics (Patton 1994). Moreover, despite a growth of studies in different areas addressing questions of risk and its social context, the dominant paradigm of 'risk groups' and individual behaviours continues to be a key focus (Craddock 2000). The concepts of 'risk group' and 'risk behaviour' are well-established in public health research as well as by geographers to identify the determinants of healthy lifestyles as well as chronic ill health. The nature and intensity of risk varies by gender, race, ethnicity and class and, in addition, the risk group idea moves in the direction of peopling the virus (Kearns 1996). In developing the 'new geographies of HIV/AIDS' which are alternatives to the quantitative approach, many authors (i.e., Downing 2008; Gilbert and Walker 2009; Ghosh et al. 2009; Kuhanen 2010; Schatz 2011; Seckinelgin 2012; Campbell et al. 2012; Tobin et al. 2013; Lewis 2016; Collins et al. 2016) highlighted different issues such as women vulnerability, complex management of HIV/AIDS treatment, economic disparities, sexual culture change, implications of the HIV research, biomedical and behavioural models, place and social networks, urban ecological approach, territorial stigma, etc. through insightful narratives.

4. HIV/AIDS literature in Bangladesh and research gaps

The HIV epidemic has brought new dimensions of risk to third world countries like Bangladesh. Along with behavioural and biomedical risk, Bangladesh has many contextual features, including poverty, gender inequality, stigma and discrimination, etc. that are relevant to HIV risk and vulnerabilities. It has brought many changes, not only in the health sector, but also in social and economic issues. These changes have occurred in the everyday lives of the individuals and communities who are highly vulnerable or affected by the HIV threat. Over the last 30 years, around the world many geographers, sociologists, anthropologists, psychologists, epidemiologists and medical scientists have had an increasing interest in HIV and AIDS and its related subjects. In Bangladesh this influence has been very limited. Most of the research on HIV in Bangladesh has been performed by public health scientists and epidemiologists. Very little academic research on HIV has addressed the socio-economic or socio-geographic issues of the marginalized and stigmatized communities who are considered as the 'risk group' for HIV infection in Bangladesh. Although most epidemiological research (for example, Sarkar et al. 1998; Gibney et al. 1999a,1999b; Caldwell et al. 1999; Hosain and Chatterjee 2005, etc.) has hinted at the necessity of doing intensive study on geographically significant places like the border towns of Bangladesh, there has not been significant academic research in this area. Furthermore, there has been rare academic work on the issues of 'lifeworlds', identity and stigma, risk behaviour and coping of the marginalized people in HIV discourse such as commercial sex workers, drug users and HIV positive people in conjunction with HIV policy issues in Bangladesh. The following section shows the main academic research trend or emphasis on HIV-related issues in Bangladesh and the current gaps in research.

There is a vast literature on epidemiology and aetiology for the most at-risk groups (Azim et al.2000,2004; Shirin et al. 2000; Rahman and Zaman 2005; Bosu 2013, etc.). These reports have looked not only at HIV infection but also at STDs and hepatitis where a significant number of works took place in Dhaka, particularly on IDU (Intravenous Drug Users) and transport workers (for example, Roy et al. 2010). Among sex workers, many research projects are carried out with brothel-based sex workers (for example, Jenkins 1999; Nessa et al. 2005 etc.) and blood donors (Mollah et al. 2004). Very little work (for example, Haseen et al. 2012) has been covered on residence and hotel girls around the country. Some research (for example, Bloem et al. 1999; Motiur Rahman et al. 2000 etc.) can be seen in relation to street girls but most is based in Dhaka. Most public health work (for example, Sabin 1998; Gibney et al. 2001; Islam et al. 2003; Nessa et al. 2004; Azim et al. 2008; Hossain et al. 2012; Mowla and Sattar 2016etc.) has assessed risk through blood testing. Secondly, much research (for example, Caldwell and Pieris 1999; Gibney et al. 2003; Chan and Khan 2007; Alam et al. 2013; Kamal et al. 2015 etc.) has assessed risk behaviours for HIV prevalence of sex workers and drug users as well as truckers. These previously mentioned studies have tried to show the potentiality of the future HIV threat to Bangladesh from the behavioural as well as the biomedical point of view. Thirdly, there are a good number of papers (for example, Rahman et al. 1999; Khan 2002; Islam et al. 2002; Sarafian 2012 etc.) which have focused on the assessment of awareness of HIV/AIDS among the different stakeholders, especially the most at-risk groups for HIV. Finally, all the previous categories of work discuss some relevant preventive efforts in their concluding remarks. With a dominant biomedical and epidemiological framework, many studies of HIV in Bangladesh frequently ignore the explanatory issues regarding the prejudice against marginalized and vulnerable people, health risks, unhealthy places and policy, all of which can be investigated by utilizing qualitative methodologies. Following Mann (1987), epidemiological data alone cannot represent the multiple and complex social dimensions of HIV/AIDS. According to him, the pandemic has been conceptualized as consisting of three separate phases: an epidemic of HIV infection, an epidemic of AIDS and an epidemic of social, cultural, economic and political responses to AIDS. The third of these has been the most explosive, characterized by denial, stigma and discrimination. However, the substantial literature supports the notion that the HIV/AIDS epidemic is determined by a combination of structural, social/cultural and individual factors (UNAIDS 1997; Parker et al. 2000; Scambler and Paoli 2008). In the following, the author divides the existing literature gap on HIV in Bangladesh into five categories: 'lifeworlds' issue; identity and stigma; risk behaviour and coping; place concern; and policy and practice.

4.1 'Lifeworlds' issue

Apart from a few studies assessing the prevalence of STDs/HIV, only a few academic research works have addressed the 'lifeworlds' of marginalized people in Bangladesh, such as sex workers and drug users. Most work has tried to cover the lifestyle of a particular group, but most has only incomplete views of their whole lives. Among the works, Khan and Arefeen (1992) and Ahsan et al. (1999) are notable. In reviewing the existing literature, it can be seen that there is a lack of information on social issues of HIV-vulnerable people. A small number of newspaper articles point out some social problems, but there is no detailed or in-depth work into how these social problems are related to economic uncertainty and daily vulnerability (Paul 2009). The 'lifeworlds' of marginalized communities in terms of their everyday practices or customs, emotions and aspirations, monetary uncertainty and everyday suffering, anger and expectations from society can be explored through qualitative approach such as in-depth interviewing, focus group discussion and naturalistic observation, etc.

4.2 Identity and stigma

There is much research on sex workers' and drug users' risk of HIV in Bangladesh, but no detailed work could be discovered on their social status or the stigmatization or rights issue. A limited literature has addressed the role of social, cultural and behavioural practices in explaining marginalization and vulnerability of HIV. Although there are a few recent works on sex workers' identity (Chowdhury 2006), rights (Ara 2005) and harassment and suffering (Hasan 2007; Ullah 2005,2011; Hossain and Kippax 2011), no major academic works have approached drug users' and HIV positives' marginalization and stigmatization except a Human Rights Watch Report (2003). There is no information on how HIV-affected or HIV-prone people in Bangladesh live with social hazards. Still less research in Bangladesh has focused on the way in which these social and psychological factors affect people's everyday lives, except for Paul (2009). He was devoted to the task of studying the sociogeographical and psychological aspects of having HIV or living in close proximity to people with HIV through empirical research.

4.3 Risk behaviour and coping

There are rich descriptions of HIV risk behaviours, including behavioural and biomedical approaches to HIV transmission (for example, Alam et al. 2013; Karim and Mona 2013; Saha and Nari Mukti Sangha 2015, etc.), but there is a serious lack of qualitative information on the pattern and nature of risk in Bangladesh. There is a gap in the literature on coping strategies of vulnerable people to HIV infection and risk. 'Risk' should be discussed in light of the implications for understanding how everyday norms influence the ways in which people perceive risk and act in response to risk (please see Paul 2009). There should be an examination of how 'risk' is perceived among sex workers, drug users and transport workers, and how their 'norms' influence individual attempts at 'risk behaviour' change, as coping strategies, by utilizing qualitative methods.

4.4 Place concern

The Bangladeshi academic literature ignores the role and importance of place in 'contributing' to health risks, particularly HIV. Some publications (for example, Islam et al.1999) have alarmed public health officials about the dramatic rise of HIV infection in Bangladesh's neighbouring countries, especially in India. But there is an absence of major work (except Caldwell et al. 1999; Khan 2003; Gazi et al. 2008) examining the role of significant places like borders, people trafficking zones and port areas in transmitting the potential source of HIV infection in Bangladesh. Apart from sexual networking and its related impact, there is no research on place as a source of stigma for vulnerable people and their health risks. Again, very little research has been done on the changing condition of brothels and the impact on transmission-related risk, except for Jenkins and Rahman (2002). There has also been very little research on destitute women or slum women, except for Gibney et al. (2001) and Sabin's (1998) work. Similarly, with regard to border girls, there is little research (except Paul and Hasnath 2000; Paul et al. 2012) on their role in the transmission of HIV. It is notable that many NGO reports can be found on the issue of women trafficking as a whole but not on the trafficked girls themselves. There is vast scope (for example, Paul 2009) to look at the probability of HIV risk for Bangladesh through trafficking and other unidentified sources.

4.5 Policy and practice

Policies and institutional practices can make certain populations vulnerable to disease. Specifically, poverty, gender disparity and discrimination have been identified as structural factors facilitating health risk particularly HIV transmission worldwide (Parker et al. 2000). However, social and cultural factors, including socio-economic status, norms, values, beliefs and ideals of a population, are often intertwined with health behaviour among a population. A number of publications in Bangladesh discuss different HIV prevention measures (for example, Sarkar et al. 1998; Hawkes 2001; Sarma and Oliveras 2011; Islam et al. 2013,etc.) but there is no information on the 'contradictory programmes' for HIV prevention in Bangladesh, 'standard awareness campaigns', stigma and discrimination for HIV positives, NGO politics and donor policy and the role of the government. Very little detailed research has explored HIV policy, policy implementation or the components of HIV prevention strategies in Bangladesh.

5. Concluding remarks

The global epidemic of HIV/AIDS has come under scrutiny since the 1980s and, particularly, from the beginning of the 1990s as a substantial public health problem. As a topic of research, it has a strong pull because of the heavy toll of morbidity and mortality and also because of the intellectual challenge of understanding its epidemiology. Like many other infectious diseases, geographers have been making contributions to the 'geography of HIV/AIDS' in the contemporary period. Using quantitative approaches, many medical geographers have contributed to the distribution and diffusion of the HIV research across the world. Analysis of the geographic distribution and migration of HIV is an established field of study where mapping is the main element to show the origin of the virus and its diffusion over space. They have explored questions pertaining to the origin, aetiology and spatial diffusion of the disease by using available hypotheses, theories and data following epidemiological approach. But many health geographers have considered that there has been a strange silence in the geographical literature about the investigation of risk and its geographic variables in the discourse of HIV/AIDS. In other words, many health geographers have recognized the socio-political or economic dimensions of the HIV-related sufferings or vulnerabilities, following a qualitative method. In this way, geographers have been constructing a 'new geographies of HIV/AIDS', which are alternatives to the epidemiological approach. The HIV epidemic has brought many challenges of health and social risk to third world countries like Bangladesh which ultimately hamper the everyday lives of the individuals and communities who are highly vulnerable or affected by the HIV threat. Although many geographers around the world over the last three decades have paid attention to the field of HIV and AIDS, this influence has been very limited in Bangladesh and many arenas of research in the HIV field have not been covered equally. In this review, it is seen that most HIV-related research is based on epidemiological and public health perspectives and there is a little in-depth research concerning the socio-cultural and geographic impacts of the HIV disaster in Bangladesh. With a dominant biomedical and epidemiological framework, many studies of HIV in Bangladesh frequently ignore the explanatory issues regarding the prejudice against marginalized and vulnerable people, health risks, unhealthy places and policy, all of which can be investigated by utilizing qualitative methodologies.

Almost all of the literature shows HIV as an epidemiological problem rather than investigating it from a social or cultural or geographical point of view. There has been rare academic work on the issues of 'lifeworlds', identity and stigma, risk behaviour and coping of marginalized people such as commercial sex workers, drug users and HIV positive people in HIV discourse. Moreover, there has not been significant academic research on geographically significant places like the border towns of Bangladesh, though most epidemiological research has hinted at the necessity for doing intensive research. The present chapter has focused on the main academic research trend or emphasis of HIV-related issues in Bangladesh and the current gaps in research. This chapter has explored the global perspectives of health geography literature related to health risk, particularly HIV. In reviewing the geographic and HIV literatures of Bangladesh from local contexts, the author has found a conceptual and theoretical (including methodological) framework for developing its research gaps. The qualitative focus in HIV research among marginalized and vulnerable people can be aimed to investigate their 'risk' and 'vulnerability' through their 'lifeworlds', stigmatized identity, discrimination and criminalization. The investigation will be advantageous in measuring the health risk from social and geographical points of view. Furthermore, this chapter has explored some of these research gaps in the context of health risks and also investigated how HIV-vulnerable people can benefit from better planning for mitigation and prevention.

Acknowledgements

This chapter is based on the author's doctoral study conducted at the Department of Geography, University of Durham, UK in 2009. The author is very grateful to Professor Dr Peter J Atkins and Dr Christine E Dunn of the above department for their supervision during the PhD programme. The author is also grateful to the authority of ORSAS/UK and doctoral fellowship programme of Durham University for their financial awards.

References

Aase, A. and S. Agyei-Mensah. 2005. 'HIV/AIDS in Sub-Saharan Africa: Geographical Perspectives'. *Norsk Geografisk Tidsskrift-Norwegian Journal of Geography*, 59(1): 1–5.

Ahsan, R. M., N. Ahmad, A. Z. Eusuf and J. Roy. 1999. 'Prostitutes and their Environment in Narayanganj, Bangladesh'. Asia *Pacific Viewpoint*, 40(1): 33–44.

Alam, N., M. E. Chowdhury, M. K. Mridha, A. Ahmed, L. J. Reichenbach, P. K. Streatfield and T. Azim. 2013. 'Factors Associated With Condom Use Negotiation by Female Sex Workers in Bangladesh'. *International Journal of STD & AIDS*, 24(10): 813–821.

Alonzo, A. A. and N. R. Reynolds.1995. 'Stigma, HIV and AIDS: An Exploration and Elaboration of a Stigma Trajectory'. *Social Science & Medicine*, 41(3): 303–315.

Ara, Z. 2005. 'Violation &Denial of Access to Health-Rights for Women Involved in Commercial Sex Work in Bangladesh'. *Women's Health and Urban Life*, 4(1): 6–26.

Asthana, S. 1998. 'The Relevance of Place in HIV Transmission and Prevention: The Commercial Sex Industry in Madras'. In R. A. Kearns and W. M. Gesler (eds.), *Putting Health Into Place: Landscape, Identity, and Wellbeing*, pp. 168–190. Syracuse: Syracuse University Press.

Asthana, S. and R. Oostvogels. 1996. 'Community Participation in HIV Prevention: Problems and Prospects for Community-based Strategies Among Female Sex Workers in Madras'. *Social Science and Medicine*, 43(2): 133–148.

Azim, T., M. S. Alam, M. Rahman, M. S. Sarker, G. Ahmed, M. R. Khan, S. Rahman, A. S. M. M. Rahman and D. A. Sack. 2004. 'Impending Concentrated HIV Epidemic Among Injecting Drug Users in Central Bangladesh'. *International Journal of STD & AIDS*, 15(4): 280–282.

Azim, T., M. N. Islam, J. Bogaerts, M. A. H. Mian, M. S. Sarker, K. R. Fattah, P. Simmonds, C. Jenkins, M. R. Choudhury and V. I. Mathan.2000. 'Prevalence of HIV and Syphilis Among High-Risk Groups in Bangladesh'. *AIDS*, 14(2): 210–211.

Azim, T., M. Rahman, M. S. Alam, I. A. Chowdhury, R. Khan, M. Reza, M. Rahman, E. I. Chowdhury, M. Hanifuddin and A. S. M. M. Rahman. 2008. 'Bangladesh Moves From Being a Low-Prevalence Nation for HIV to One With a Concentrated Epidemic in Injecting Drug Users'. *International Journal of STD & AIDS*, 19(5): 327–331.

Bloem, M., E. Hoque, L. Khanam, T. S. Mahbub, M. Salehin and S. Begum. 1999. 'HIV/AIDS and Female Street-based Sex Workers in Dhaka City: What About Their Clients?'. In J. C. Caldwell, P. Caldwell, J. Anarfi et al. (eds.), *Resistance to Behavioural Change to Reduce HIV/AIDS Infection in Predominantly Heterosexual Epidemics in Third World Countries*, pp. 197–210. Canberra: Health Transition Centre, Australian National University.

Bosu, A. 2013. 'Sexual Act With Multiple Sex Partners Is an Upcoming Threat for Rapid Transmission of HIV Among People Who Inject Drugs (PWID) in Bangladesh'. *Sexually Transmitted Infections*, 89(Suppl 1): A179–A179.

Brown, M. 1995. 'Ironies of Distance: An Ongoing Critique of the Geographies of AIDS'. *Environment and Planning D: Society and Space*, 13(2): 159–183.

Brown, M. 1997. *Replacing Citizenship: AIDS Activism and Radical Democracy*. London: Guilford Press.
Brown, M. 2012. 'Gender and Sexuality I: Intersectional Anxieties'. *Progress in Human Geography*, 36(4): 541–550.
Browne, A. W. and H. R. Barrett. 2001. 'Moral Boundaries: The Geography of Health Education in the Context of the HIV/AIDS Pandemic in Southern Africa'. *Geography*, 86(1): 23–36.
Caldwell, B. and I. Pieris. 1999. 'Continued High-Risk Behaviour Among Bangladeshi Males'. In J. C. Caldwell, P. Caldwell and J. Anarfi et al.(eds.), *Resistance to Behavioural Change to Reduce HIV/AIDS Infection in Predominantly Heterosexual Epidemics in Third World Countries*, pp. 183–196. Health Transition Centre, Canberra: Australian National University.
Caldwell, B., I. Pieris, Barkat-e-Khuda, J. Caldwell and P. Caldwell. 1999. 'Sexual Regimes and Sexual Networking: The Risk of an HIV/AIDS Epidemic in Bangladesh'. *Social Science and Medicine*, 48(8): 1103–1116.
Campbell, C., F. Cornish and M. Skovdal. 2012. 'Local Pain, Global Prescriptions? Using Scale to Analyse the Globalisation of the HIV/AIDS Response'. *Health & Place*, 18(3): 447–452.
Carrel, M., M. Janko, M. K. Mwandagalirwa, C. Morgan, F. Fwamba, J. Muwonga, A. K. Tshefu, S. Meshnick and M. Emch. 2016. 'Changing Spatial Patterns and Increasing Rurality of HIV Prevalence in the Democratic Republic of the Congo Between 2007 and 2013'. *Health & Place*, 39, May: 79–85.
Chan, P. A. and O. A. Khan. 2007. 'Risk Factors for HIV Infection in Males Who Have Sex With Males (MSM) in Bangladesh'. *BMC Public Health*, 7, July: 153.
Chowdhury, R. 2006. 'Outsiders and Identity Reconstruction in the Sex Workers Movement in Bangladesh'. *Sociological Spectrum*, 26(3): 335–357.
Cliff, A. D. and P. Haggett. 1988. *Atlas of Disease Distributions: Analytic Approaches to Epidemiological Data*. New York: Blackwell.
Cliff, A. D. and M. R. Smallman-Raynor. 1992. 'The AIDS Pandemic: Global Geographical Patterns and Local Spatial Processes'. *Geographical Journal*, 158(2): 182–198.
Collins, A. B., S. Parashar, K. Closson, R. B. Turje, C. Strike and R. McNeil. 2016. 'Navigating Identity, Territorial Stigma, and HIV Care Services in Vancouver, Canada: A Qualitative Study'. *Health & Place*, 40, July: 169–177.
Craddock, S. 2000. 'Disease, Social Identity, and Risk: Rethinking the Geography of AIDS'. *Transaction of the Institute of British Geographers*, 25(2): 153–168.
Del Casino Jr, V. J. 2001. 'Healthier Geographies: Mediating the Gaps Between the Needs of People Living With HIV/AIDS and Health Care in Chiang Mai, Thailand'. *The Professional Geographer*, 53(3): 407–421.
Downing Jr, M. J. 2008. 'The Role of Home in HIV/AIDS: A Visual Approach to Understanding Human-environment Interactions in the Context of Long-term Illness'. *Health & Place*, 14(2): 313–322.

Dutt, A. K., C. B. Monroe, H. M. Dutta and B. Prince.1987. 'Geographical Patterns of AIDS in the United States'. *Geographical Review*, 77(4): 456–471.
Ezekiel, M. J., A. Talle, K. S. Mnyika and K.-I. Klepp. 2010. 'Social Geography of Risk: The Role of Time, Place and Antiretroviral Therapy in Conceptions About the Sexual Risk of HIV in Kahe, Kilimanjaro, Tanzania'. *Norwegian Journal of Geography*, 64(1): 48–57.
Feldacker, C., M. Emch and S. Ennett. 2010. 'The Who and Where of HIV in Rural Malawi: Exploring the Effects of Person and Place on Individual HIV Status'. *Health & Place*, 16(5): 996–1006.
Ferguson, A. G. and C. N. Morris. 2007. 'Mapping Transactional Sex on the Northern Corridor Highway in Kenya'. *Health & Place*, 13(2): 504–519.
Gallo, R. C. 1987. 'The AIDS Virus'. *Scientific American*, 256(1): 39–48.
Gazi, R., A. Mercer et al. 2008. 'An Assessment of Vulnerability to HIV Infection of Boatmen in Teknaf, Bangladesh'. *Conflict and Health*, 2(5): 1–11.
Gesler, W. M. 2006. 'Geography of Health and Healthcare'. In B. Warf (ed.), *Encyclopedia of Human Geography*, pp. 205–206. Thousand Oaks: SAGE Publications.
Gesler, W. M. and R. A. Kearns. 2002. *Culture/Place/Health*. London: Routledge.
Ghosh, J., V. Wadhwa and E. Kalipeni. 2009. 'Vulnerability to HIV/AIDS Among Women of Reproductive Age in the Slums of Delhi and Hyderabad, India'. *Social Science & Medicine*, 68(4): 638–642.
Gibney, L., P. Choudhury, Z. Khawaja, M. Sarker and S. H. Vermund.1999a. 'Behavioural Risk Factors for HIV/AIDS in a Low-HIV Prevalence Muslim Nation, Bangladesh'. *International Journal of STD & AIDS*, 10(3): 186–194.
Gibney, L., P. Choudhury, Z. Khawaja, M. Sarker and S. H. Vermund. 1999b. 'HIV/AIDS in Bangladesh: An Assessment of Biomedical Risk Factors for Transmission'. *International Journal of STD & AIDS*, 10(5): 338–346.
Gibney, L., M. Macaluso, K. Kirk, M. S. Hassan, J. Schwebke, S. H. Vermund and P. Choudhury. 2001. 'Prevalence of Infectious Diseases in Bangladeshi Women Living Adjacent to a Truck Stand: HIV/STD/Hepatitis/Genital Tract Infections'. *Sexually Transmitted Infections*, 77(5): 344–350.
Gibney, L., N. Saquib and J. Metzger. 2003. 'Behavioural Risk Factors for STD/HIV Transmission in Bangladesh's Trucking Industry'. *Social Science and Medicine*, 56(7): 1411–1424.
Gilbert, L. and L. Walker. 2009. 'They (ARVs) Are My Life, Without Them I'm Nothing'– Experiences of Patients Attending a HIV/AIDS Clinic in Johannesburg, South Africa'. *Health & Place*, 15(4): 1123–1129.
Gould, P.1989. 'Geographic Dimensions of the AIDS Epidemic'. *The Professional Geographer*, 41(1): 71–78.
Gould, P.1993. *The Slow Plague: Geography of the AIDS Pandemic*. Oxford: Blackwell.
Gould, P. and R. Wallace. 1994. 'Spatial Structures and Scientific Paradoxes in the AIDS Pandemic'. *Geografiska Annaler*, 76(2): 105–116.

Hasan, M. 2007. *Harassment Pattern of Sex Workers in Bangladesh: A Situational Analysis of Three Brothels, Narratives and Perspective in Sociology Understanding the Past, Envisaging the Future.* Proceedings of the 8th Annual Conference of Hong Kong Sociological Association, Hong Kong Shue Yan University.

Haseen, F., F. A. H. Chawdhury, M. E. Hossain, M. Huq, M. U. Bhuiyan, H. Imam, D. M. M. Rahman, R. Gazi, S. I. Khan, R. Kelly and J. Ahmed. 2012. 'Sexually Transmitted Infections and Sexual Behaviour Among Youth Clients of Hotel-Based Female Sex Workers in Dhaka, Bangladesh'. *International Journal of STD & AIDS*, 23(8): 553–559.

Hawkes, S. 2001. 'Human Immunodeficiency Virus and Hepatitis in Bangladesh: Widespread or Targeted Prevention Strategies? Commentary'. *International Journal of Epidemiology*, 30(4): 885–886.

Hosain, G. M. M. and N. Chatterjee. 2005. 'Beliefs, Sexual Behaviours and Preventive Practices With Respect to HIV/AIDS Among Commercial Sex Workers in Daulatdia, Bangladesh'. *Public Health*, 119(5): 371–381.

Hossain, K. J., N. Akter, M. Kamal, M. C. Mandal and Md Aktharuzzaman. 2012. 'Screening for HIV Among Substance Users Undergoing Detoxification'. *International Journal of STD & AIDS*, 23(10): e1–e5.

Hossain, M. B. and S. Kippax. 2011. 'Stigmatized Attitudes Toward People Living With HIV in Bangladesh: Health Care Workers' Perspectives'. *Asia Pacific Journal of Public Health*, 23(2): 171–182.

Human Rights Watch Report. 2003. *Ravaging the Vulnerable: Abuse Against Persons at High Risk of HIV Infection*. August 19, USA.

Islam, M. S., S. Rasin and L. Rahman. 2013. 'Antiretroviral Therapy (ART) Programme in Bangladesh: Increasing National Ownership'. *Sexually Transmitted Infections*, 89(Suppl 1): A378–A378.

Islam, M. T., A. K. Mitra, A. H. Mian and S. H. Vermund. 1999. 'HIV/AIDS in Bangladesh: A National Surveillance'. *International Journal of STD & AIDS*, 10(7): 471–474.

Islam, M. T., G. Mostafa, A. U. Bhuiya, S. Hawkes and A. de Francisco. 2002. 'Knowledge on, and Attitude Toward, HIV/AIDS Among Staff of an International Organization in Bangladesh'. *Journal of Health, Population and Nutrition*, 20(3): 271–278.

Islam, S. K., K. J. Hossain, M. Kamal and M. Ahsan. 2003. 'Prevalence of HIV Infection in the Drug Addicts of Bangladesh: Drug Habit, Sexual Practice and Lifestyle'. *International Journal of STD & AIDS*, 14(11): 762–764.

Jenkins, C. 1999. 'Resistance to Condom Use in a Bangladesh Brothel'. In J. C. Caldwell, P. Caldwell, J. Anarfi et al.(eds.), *Resistance to Behavioural Change to Reduce HIV/AIDS Infection in Predominantly Heterosexual Epidemics in Third World Countries*, pp. 211–222. Health Transition Centre, Canberra: Australian National University.

Jenkins, C. and H. Rahman. 2002. 'Rapidly Changing Conditions in the Brothels of Bangladesh: Impact on HIV/STD'. *AIDS Education and Prevention*, 14(Suppl A): 97–106.

Kalipeni, E., S. Craddock, J. R. Oppong and J. Ghosh (eds.). 2004. *HIV and AIDS in Africa: Beyond Epidemiology*. Oxford: Blackwell.

Kalipeni, E. and L. Zulu.2008. 'Using GIS to Model and Forecast HIV/AIDS Rates in Africa'. 1986-2010, *The Professional Geographer*, 60(1): 33–53.

Kamal, S. M., C. H. Hassan and R. H. Salikon. 2015. 'Safer Sex Negotiation and Its Association With Condom Use Among Clients of Female Sex Workers in Bangladesh'. *Asia Pacific Journal of Public Health*, 27(2): NP2410–NP2422.

Karim, M. and N. Mona.2013. 'Knowledge Attitude Practises About Sexually Transmitted Disease Among the Commercial Sex Workers'. *Sexually Transmitted Infections*, 89(Suppl 1): A204–A204.

Kearns, R. A. 1996. 'AIDS and Medical Geography: Embracing the Other?'. *Progress in Human Geography*, 20(1):123–131.

Kearns, R. A. and G. Moon. 2002. 'From Medical to Health Geography: Novelty, Place and Theory After a Decade of Change'. *Progress in Human Geography*, 26(5): 605–625.

Kesby, M. 2000. 'Participatory Diagramming: Deploying Qualitative Methods Through an Action Research Epistemology'. *Area*, 32(4): 423–435.

Khan, J. 2003. 'Spatiality of Truckers Mobility and Potential STD/AIDS Incidence in Bangladesh'. *Journal of BNGA*, 31(1): 62–71.

Khan, M. A. 2002. 'Knowledge on AIDS Among Female Adolescents in Bangladesh: Evidence From the Bangladesh Demographic and Health Survey Data'. *Journal of Health, Population and Nutrition*, 20(2): 130–137.

Khan, Z. R. and H. K. Arefeen. 1992. *Prostitution in Bangladesh*. Dhaka: Centre for Social Studies, Mimeographed Monograph.

Kuhanen, J. 2010. 'Sexualised Space, Sexual Networking & the Emergence of AIDS in Rakai, Uganda'. *Health & Place*, 16(2): 226–235.

Lam, N. S. N. and K. B. Liu.1996. 'Use of Space-Filling Curves in Generating a National Rural Sampling Frame for HIV/AIDS Research'. *The Professional Geographer*, 48(3): 321–332.

Lewis, N. M. 2016. 'Urban Encounters and Sexual Health Among Gay and Bisexual Immigrant Men: Perspectives From the Settlement and Aids Service Sectors'. *Geographical Review*, 106(2): 235–256.

Lindquist, J. 2005. 'Organizing AIDS in the Borderless World: A Case Study From the Indonesia-Malaysia-Singapore Growth Triangle'. *Asia Pacific Viewpoint*, 46(1): 49–63.

Madise, N. J., A. K. Ziraba, J. Inungu, S. A. Khamadi, A. Ezeh, E. M. Zulu, J. Kebaso, V. Okoth and M. Mwau. 2012. 'Are Slum Dwellers at Heightened Risk of HIV Infection Than Other Urban Residents? Evidence From Population-based HIV Prevalence Surveys in Kenya'. *Health & Place*, 18(5): 1144–1152.

Magadi, M. and M. Desta. 2011. 'A Multilevel Analysis of the Determinants and Cross-national Variations of HIV Seropositivity in Sub-Saharan Africa: Evidence From the DHS'. *Health & Place*, 17(5): 1067–1083.

Mann, J. M. 1987. 'The Global AIDS Situation'. *World Health Statistics Quarterly*, 40(2): 185–192.

Marten, L. 2005. 'Commercial Sex Workers: Victims, Vectors or Fighters of the HIV Epidemic in Cambodia?'. *Asia Pacific Viewpoint*, 46(1): 21–34.
May, J. M. 1958. *The Ecology of Human Disease*. New York: MD Publication.
Mayer, J. D. 2005. 'The Geographical Understanding of HIV/AIDS in Sub-Saharan Africa'. *Norwegian Journal of Geography*, 59(1): 6–13.
Mollah, A. H., M. A. Siddiqui, K. S. Anwar, F. J. Rabbi, Y. Tahera, Md S. Hassan and N. Nahar. 2004. 'Seroprevalence of Common Transfusion-Transmitted Infections Among Blood Donors in Bangladesh'. *Public Health*, 118(4): 299–302.
Moran, D. 2005. 'The Geography of HIV/AIDS in Russia: Risk and Vulnerability in Transition'. *Eurasian Geography and Economics*, 46(7): 525–551.
Mowla, M. R. and M. A. Sattar. 2016. 'Recent Trends in Sexually Transmitted Infections: The Chittagong, Bangladesh Experience'. *Sexually Transmitted Infections*, 92(5): 349–349.
Nessa, K., S. A. Waris, A. Alam, M. Huq, S. Nahar, F. A. Hasan Chawdhury, S. Monira et al.2005. 'Sexually Transmitted Infections Among Brothel-based Sex Workers in Bangladesh: High Prevalence of Asymptomatic Infection'. *Sexually Transmitted Diseases*, 32(1): 13–19.
Nessa, K., S. A. Waris, Z. Sultan, S. Monira et al. 2004. 'Epidemiology and Etiology of Sexually Transmitted Infection Among Hotel-based Sex Workers in Dhaka, Bangladesh'. *Journal of Clinical Microbiology*, 42(2): 618–621.
Oppong, J. R. 1998. 'A Vulnerability Interpretation of the Geography of HIV/AIDS in Ghana, 1986–1995'. *The Professional Geographer*, 50(4): 437–448.
Oppong, J. R. and A. Harold. 2009. 'Disease, Ecology, and Environment'. In T. Brown, S. McLafferty and G. Moon (eds.), *A Companion to Health and Medical Geography*, pp. 81–95. London: Blackwell-Wiley.
Parker, R. G., D. Easton and C. H. Klein. 2000. 'Structural Barriers and Facilitators in HIV Prevention: A Review of International Research'. *AIDS*, 14(Suppl 1): S22–32.
Patton, C. 1994. *Last Served? Gendering the HIV Pandemic*. London: Taylor & Francis.
Paul, A. 2009. *Geographies of HIV/AIDS in Bangladesh: Vulnerability, Stigma and Place*. Durham Theses, Durham University. http://etheses.dur.ac.uk/1348/ (last accessed on 3rd December 2018).
Paul, A., P. J. Atkins and C. E. Dunn. 2012. 'Borders and HIV Risk: A Qualitative Investigation in Bangladesh'. *Oriental Geographer*, 53(1): 73–82.
Paul, B. K. 1994. 'AIDS in Asia'. *Geographical Review*, 84(4): 367–379.
Paul, B. K. and S. A. Hasnath. 2000. 'Trafficking in Bangladeshi Women and Girls'. *Geographical Review*, 90(2): 268–276.
Piper, N. and B. S. A. Yeoh. 2005. 'Introduction: Meeting the Challenges of HIV/AIDS in Southeast and East Asia'. *Asia Pacific Viewpoint*, 46(1): 1–5.
Pyle, G. F. 1969. 'The Diffusion of Cholera in the United States in the Nineteenth Century'. *Geographical Analysis*, 1(1): 59–75.
Rahman, M., A. Alam, K. Nessa, A. Hossain et al. 2000. 'Etiology of Sexually Transmitted Infections Among Street-based Female Sex Workers in Dhaka, Bangladesh'. *Journal of Clinical Microbiology*, 38(3): 1244–1246.

Rahman, M., T. A. Shimu, T. Fukui, T. Shimbo and W. Yamamoto. 1999. 'Knowledge, Attitudes, Beliefs and Practices About HIV/AIDS Among the Overseas Job Seekers in Bangladesh'. *Public Health*, 113(1): 35–38.

Rahman, M. and M. S. Zaman. 2005. 'Awareness of HIV/AIDS and Risky Sexual Behaviour Among Male Drug Users of Higher Socioeconomic Status in Dhaka, Bangladesh'. *Journal of Health, Population and Nutrition*, 23(3): 298–301.

Ransome, Y., I. Kawachi, S. Braunstein and D. Nash. 2016. 'Structural Inequalities Drive Late HIV Diagnosis: The Role of Black Racial Concentration, Income Inequality, Socioeconomic Deprivation, and HIV Testing'. *Health & Place*, 42, November: 148–158.

Rosenberg, M. W. 1998. 'Medical or Health Geography? Populations, People and Places'. *International Journal of Population Geography*, 4(3): 211–226.

Roy, T., C. Anderson, C. Evans and M. S. Rahman. 2010. 'Sexual Risk Behaviour of Rural-to-Urban Migrant Taxi Drivers in Dhaka, Bangladesh: A Cross-sectional Behavioural Survey'. *Public Health*, 124(11): 648–658.

Sabatier, R. 1996. 'Migrants and AIDS: Themes of Vulnerability and Resistance'. In M. Haour-Knipe and R. Rector (eds.), *Crossing Borders: Migration, Ethnicity and AIDS*. London: Taylor & Francis.

Sabin, K. M. 1998. *A Study of Sexually Transmitted Infections and Associated Factors in Slum Communities of Dhaka, Bangladesh*. PhD Dissertation, The School of Hygiene and Public Health, The Johns Hopkins University, Baltimore, Maryland.

Saha, T. K. and Nari Mukti Sangha. 2015. 'To Prevent HIV/AIDS Through Awareness Raising and Social Behaviours Change of Sex Worker in Bangladesh: An Experience From Kandapara Brothel, Tangail, Bangladesh'. *Sexually Transmitted Infections*, 91(Suppl 2): A206.

Sarafian, I. 2012. 'Process Assessment of a Peer Education Programme for HIV Prevention Among Sex Workers in Dhaka, Bangladesh: A Social Support Framework'. *Social Science & Medicine*, 75(4): 668–675.

Sarkar, S., N. Islam, F. Durandin et al. 1998. 'Low HIV and High STD Among Commercial Sex Workers in a Brothel in Bangladesh: Scope for Prevention of Larger Epidemic'. *International Journal of STD & AIDS*, 9(1): 45–47.

Sarma, H. and E. Oliveras. 2011. 'Improving STI Counselling Services of Non-Formal Providers in Bangladesh: Testing an Alternative Strategy'. *Sexually Transmitted Infections*, 87(6): 476–478.

Scambler, G. and F. Paoli. 2008. 'Health Work, Female Sex Workers and HIV/AIDS: Global and Local Dimensions of Stigma and Deviance as Barriers to Effective Interventions'. *Social Science & Medicine*, 66(8): 1848–1862.

Schatz, E., S. Madhavan and J. Williams. 2011. 'Female-headed Households Contending With AIDS-related Hardship in Rural South Africa'. *Health & Place*, 17(2): 598–605.

Seckinelgin, H. 2012. 'The Global Governance of Success in HIV/AIDS Policy: Emergency Action, Everyday Lives and Sen's Capabilities'. *Health & Place*, 18(3): 453–460.

Shannon, G. W. and G. F. Pyle. 1989. 'The Origin and Diffusion of AIDS: A View From Medical Geography'. *Annals of the Association of American Geographers*, 79(1): 1–24.

Shirin, T., T. Ahmed, A. Iqbal, M. Islam and M. N. Islam. 2000. 'Prevalence and Risk Factors of Hepatitis B Virus, Hepatitis C Virus, and Human Immunodeficiency Virus Infections Among Drug Addicts in Bangladesh'. *Journal of Health, Population and Nutrition*, 18(3): 145–150.

Smallman-Raynor, M. R. and A. D. Cliff. 1990. 'Acquired Immune Deficiency Syndrome (AIDS): Literature, Geographical Origins and Global Patterns'. *Progress in Human Geography*, 14(2): 157–213.

Stamp, L. D. 1964. *The Geography of Life and Death*. Ithaca: Cornell University Press.

Takahashi, L. M. 1997. 'Stigmatization, HIV/AIDS, and Communities of Colour: Exploring Response to Human Service Facilities'. *Health & Place*, 3(3): 187–199.

Takahashi, L. M. and G. Smutny. 2001. 'Explaining Access to Human Services: The Influence of Descriptive and Behavioural Variables'. *The Professional Geographer*, 53(1): 12–31.

Teye, J. K. 2005. 'Condom Use as a Means of HIV/AIDS Prevention and Fertility Control Among the Krobos of Ghana'. *Norwegian Journal of Geography*, 59(1): 65–73.

Tobin, K. E., M. Cutchin, C. A. Latkin and L. M. Takahashi. 2013. 'Social Geographies of African American Men Who Have Sex With Men (MSM): A Qualitative Exploration of the Social, Spatial and Temporal Context of HIV Risk in Baltimore, Maryland'. *Health & Place*, 22, July: 1–6.

Ullah, A. K. M. A. 2005. 'Prostitution in Bangladesh: An Empirical Profile of Sex Workers'. *Journal of International Women's Studies*, 7(2): 111–122.

Ullah, A. K. M. A. 2011. 'HIV/AIDS-related Stigma and Discrimination: A Study of Health Care Providers in Bangladesh'. *Journal of the International Association of Physicians in AIDS Care*, 10(2): 97–104.

UNAIDS. 1997. *Impact of HIV and Sexual Health Education on the Sexual Behaviour of Young People: A Review Update*. Joint United Nations Programme on HIV/AIDS (UNAIDS), Geneva.

Valentine, G. 2007. 'Theorizing and Researching Intersectionality: A Challenge for Feminist Geography'. *The Professional Geographer*, 59(1): 10–21.

Vearey, J., I. Palmary, L. Thomas, L. Nunez and S. Drimie. 2010. 'Urban Health in Johannesburg: The Importance of Place in Understanding Intra-Urban Inequalities in a Context of Migration and HIV'. *Health & Place*, 16(4): 694–702.

Wallace, R. 1991. 'Traveling Waves of HIV Infection on a Low Dimensional "Socio-Geographic" Network'. *Social Science and Medicine*, 32(7): 847–852.

Wilton, R. D. 1996. 'Diminished Worlds? The Geography of Everyday Life With HIV/AIDS'. *Health & Place*, 2(2): 69–83.

Wilton, R. D. 1999. 'Qualitative Health Research: Negotiating Life With HIV/AIDS'. *The Professional Geographer*, 51(2): 254–264.

Wood, E., B. Yip, N. Gataric, J. S. G. Montaner, M. V. O'Shaughnessy, M. T. Schechter and R. S. Hogg. 2000. 'Determinants of Geographic Mobility Among Participants in a Population-based HIV/AIDS Drug Treatment Program'. *Health & Place*, 6(1): 33–40.

Wood, W. B. 1988. 'AIDS North and South: Diffusion Patterns of a Global Epidemic and a Research Agenda for Geographers'. *The Professional Geographer*, 40(3): 266–279.

Yao, J., V. Agadjanian and A. T. Murray. 2014. 'Spatial and Social Inequities in HIV Testing Utilization in the Context of Rapid Scale-Up of HIV/AIDS Services in Rural Mozambique'. *Health & Place*, 28, July: 133–141.

Young, L. and N. Ansell. 2003. 'Fluid Households, Complex Families: The Impacts of Children's Migration as a Response to HIV/AIDS in Southern Africa'. *The Professional Geographer*, 55(4): 464–476.

6 Geographical perspectives of scarcity and landlessness in rural households of Bangladesh

Nasreen Ahmad

1. Introduction

Land is considered as an important asset in Bangladesh. The importance has been emphasized by Hye (1996) and Rahman (1988). According to Rahman, two-thirds to about three-fourths of all assets are accounted for by land (more so for the marginal households) and Hye states that the most important cause of poverty in Bangladesh is landlessness. To Sinha (1984), landlessness is both the cause and symptom of chronic poverty. Hossain and Sen (1992), co-relating the incidence of poverty with land, also found an inverse relationship between incidence of poverty and the amount of land owned. Evidently, economic position and social linkages depend on the amount of land owned, as distribution of other assets follow the pattern of land distribution, which is sharply skewed (BBS 1989). The small/ marginal farmers do not get the benefit from new technology (irrigation) that could enhance their productivity and income generating capacity; neither do they get adequate services like safe water, sanitary latrines or have proper access to credit, education and health (Sen 1996; Ullah 1996). A positive relationship was found between economic power and social status of village families and the distribution of land or land related activities/income (Karim 1976; van Schendel 1981; Jansen 1987; Burling 1997). According to Baqee (1992), in rural areas the more land a farmer owns the more influence he exerts in the society. Apparently, land is considered to a political investment – it is invested in for status and patronage consideration (Faaland and Parkinson 1976; BRAC 1986).

Owning or not owning land is important in the context of Bangladesh for, apart from determining activity, income and social standing, the location of the land that is owned or operated or resided upon influences overall household economic condition and also effects a person's perception, attitude and responses to situations and phenomenon. Curry (1979), while mapping the famine prone thus poverty stricken area of Bangladesh, established a relationship between physical/ environmental factors and impoverishment of the rural people, while Islam (1995) holds a view that hazard prone areas (of floods/ cyclones) imbibe the people with resilience and a sense of fatalism. Landless is thus a social term given to a functional category of people who produce from land but do not own the means of control over production; people who work mainly on the land of others and have no land of their own. The International Labour Organization (ILO) defines landless (and other rural workers) as those who work in agriculture, directly and personally with no ownership rights to land, and those who, despite being owners of small marginal holdings, obtain at least 50 percent of their income as wages or payment in kind. Landlessness in Bangladesh is not a recent phenomenon and until recently, despite the importance of the concept of landlessness, there was no widely accepted and standard definition of the term. The Land Occupancy Survey (LOS) of 1977 and 1978, the national level survey on land occupancy carried out by the Bangladesh Bureau of Statistics (BBS) in collaboration with the United States Agency for International Development (USAID), provided three definitions:

- Landless 1: households with no land whatsoever
- Landless 2: those who owned only homestead but no 'other' land
- Landless 3: those who owned homestead and 0.5 acre (0.2 hectares) of 'other' land

Attempts have been made periodically both at the micro- and macro-level to assess levels, trends and causes of landlessness in Bangladesh. Presently almost all literature dealing with alleviation of rural poverty or rural development in Bangladesh devote a paragraph or two to the proportion of the population who constitute the landless in Bangladesh in general and the region in particular. Micro/village level studies mostly describe the characteristics of a landless person in terms of their needs and these data help to assess the trends of landlessness. A vast amount of literature (for example, Khan et al. 1977; BARD 1977; BARC 1978; Adnan and Rahman 1979; Murshid and Abdullah 1986; GOB 1998; Ahmad 2005, etc.) indicate a rapid increase in landlessness – at a rate faster than the population growth. The rate and level of landlessness as revealed by the studies is also not the same everywhere in the country. According to Jannuzi and Peach (1980), the landless in Bangladesh, whether residing in villages or as recent migrants in urban areas, is from the poorest of the poor, while to Adnan and Rahman (1978) the landless are the most under-privileged and exploited of the have-nots. Landless today constitutes that growing segment of the population who are often shelterless and homeless, assetless and powerless (FAO 1986).

2. Research methodology

It was observed that landlessness varied from country to country, depending on the major socio-economic variables, and this stressed the need for further study of the dynamics of the process of becoming landless. It was felt that the qualitative aspects of landlessness and the processes at work would be best captured by studies conducted over time at the household level and in a small number of areas, which are typical of the agro-ecological and other conditions in the country concerned. With this end in view, the present study is an attempt towards understanding landlessness as it operates in rural Bangladesh. Based on information of landless rural households within the various agro-ecological niches of the floodplains of Bangladesh, this study investigates, examines and analyses the spatio-temporal aspects of landlessness along with the nature, processes and the causes behind landlessness in Bangladesh.

This study is conducted in two phases. The first phase of the study was involved in the gathering of empirical background information on rural landlessness in the country by administering a simple questionnaire on heads of households in the easily accessible slum settlements within the Dhaka metropolitan area. A total of 1,000 households in 15 slum clusters from 12 thanas were randomly surveyed. The questionnaire survey conducted in the slums sought information on the household head's reasons for coming to Dhaka; the amount of land that the household possessed or had possessed; and when and how they had lost their land. Although the questionnaire survey was carried out randomly only those households that presently own some amount of land or had owned land during their 'lifetime' were enumerated. As data had been gathered from the slums of the city of Dhaka only, there was a greater representation of people coming from a few specific areas, namely, from the districts of (greater) Barisal, Faridpur, besides Dhaka and Comilla.

The second phase was conducted across the floodplains at the village level. For the main body of analysis data was gathered through questionnaire survey, discussions and observations. The floodplain was chosen for study as evidently it exhibits spatial variation of ecological conditions and thus has a varying degree of life supporting capability. It is made up of hazard prone coastal areas, waterlogged depressions and areas liable to severe flooding and bankline erosion. But above all the floodplains provide the most fertile agricultural lands of the country and support a dense population. A total of 337 households (considering a minimum of 10 percent households) in 18 villages from 18 thanas distributed over the floodplain have been surveyed for this study (Table 6.1). These households were selected on the basis of physiography, accessibility, agro-ecological variations and all belong to the landless category. Landless in this study range from those who during the time of survey owned no land whatsoever to those who owned land of 0.01 to 1.00 acre (0.004–0.4 hectares) only.

This micro-level study has adopted an approach that tends to focus on households through individuals/household heads within the village community of a particular geographical location and the causes and processes of landlessness have been analyzed through such personal/human factors as well-being, needs, capability, motivations, constraints, options and opportunities. It combines qualitative and quantitative data collected from both primary and secondary sources in its attempt at causal explanations of connections and interrelations.

3. Scarcities in the rural households of Bangladesh

Respondents of the study villages were asked to identify the times of difficulty or scarcity that households faced during a year. Irrespective of occupation and land ownership, all respondents who admitted of encountering periods of hardship during certain times of the year identified months that could be related to an agricultural season. They linked their 'hard' times to prevalent agricultural seasons, rural activity and consequent employment opportunities in their village.

3.1 Seasonal variation of scarcity

The floodplain households experiencing scarcity/hardships over the different months, as revealed through the questionnaire survey, have been presented in Table 6.2. The situation can be visualized more clearly where, with households facing scarcity in more than one month within a year, the responses have been ranked according to frequency of stated months. It may be noted that when mentioning the different months all households identified each of the Bangla calendar months separately except for Boishakh Jaistho and Ashar-Srabon – which were always stated together. This trend of response, however, helps to identify the minimum seasonal duration of household scarcity – in summer and during the rain. Empirical findings confirm that, irrespective of

Table 6.1 The study sites

Generalized ecological regions	Study village location (village/union/Thana/district)	Total households in study village	Village characteristics
Active Plain	Sarasia/Haturia-Nakalia Union/Bera Thana/Pabna	Total HH 121 Enumerated HH 13	Erosion prone Near to town Absolute landless HH 15.4 percent
Active Plain	Charkandi/Shibalaya Union/Shibalaya Thana/Manikganj	Total HH 183 Enumerated HH 18	Intense flooding Poor road network Absolute landless HH 8.8 percent
Meander Plain	Kismat Sukhenpur/Pirgacha Union/Pirgacha Thana/Rangpur	Total HH 117 Enumerated HH 12	Erosion prone No electricity Absolute landless HH 16.6 percent
Meander Plain	Chak Sadu/Gabtali Union/Gabtali Thana/Bogra	Total HH 202 Enumerated HH 20	Very fertile soil High wages Absolute landless HH 9.3 percent
Meander Plain	Nijbanail/Chandipasha Union/Nandail Thana/Mymensingh	Total HH 202 Enumerated HH 20	A three crop area High wages Absolute landless HH 10.6 percent
Meander Plain	Dakhin Doriabad/Islampur Union/Islampur Thana/Jamalpur	Total HH 300 Enumerated HH 32	Flooding area Poor work scope Absolute landless HH 13.8 percent
Estuarine Plain	Purbo Char Kali/Char Samaia Union/Bhola Sadar/Bhola	Total HH 123 Enumerated HH 12	Tidal water intrusion Landless cooperatives Absolute landless HH 22.8 percent
Estuarine Plain	Char Hasan Hossain/Char Algi Union/Ramgati Thana/Lakshmipur	Total HH 300 Enumerated HH 30	Erosion prone Poor work scope Absolute landless HH 7.9 percent
Estuarine Plain	Tatera Paschim Ram Chandrapur mouza/Dakhin Barera Union/Chandina Thana/Comilla	Total HH 300 Enumerated HH 30	Sufficient work Poor wages Absolute landless HH 6.6 percent

Region	Location	Households	Characteristics
Moribund/Deltaic Plain	Baribanka/Buripota Union/ Meherpur Thana/Meherpur	Total HH 180 Enumerated HH 18	Dry region Poor wages Absolute landless HH 8.6 percent
Moribund/Deltaic Plain	Dorisholai/Arpara Union/ Shalikha Thana/Magura	Total HH 195 Enumerated HH 20	Low areas No females in field Absolute landless HH 3.7 percent
Tidal Plain	Talbaria Alipur Mouza/Alipur Union/Satkhira Sadar Thana/ Satkhira	Total HH 145 Enumerated HH 14	Undeveloped area Poor yield for salinity Absolute landless HH 7.6 percent
Tidal Plain	Takia/Gadaipur Union/Paikgacha Thana/Khulna	Total HH 132 Enumerated HH 15	Bigger landholdings Lot of 'ghers' Absolute landless HH 12.4 percent
Coastal Plain	Purbo Habilas Dwip/Habilas Dwip Union/Patiya Thana/ Chittagong	Total HH 120 Enumerated HH 14	Known as 'ideal gram' Strong urban links Absolute landless HH 9.5 percent
Floodplain Basins/ Depressions	Modhupur-Sholobhagi/ Mollapara Union/Sunamganj Sadar Thana/Sunamganj	Total HH 150 Enumerated HH 15	Flash flood No work Estimated landless HH 17.2 percent
Floodplain Basins/ Depressions	Dakhin Roghunathpur/ Roghunathpur Union/ Gopalganj sadar Thana/ Gopalganj	Total HH 180 Enumerated HH 18	Flood prone Waterlogging Absolute landless HH 5.9 percent
Piedmont Plain (Himalayan-Northern Piedmont Plain)	Buraburi Mondolpara/Buraburi Union/Tetulia Thana/ Panchagarh	Total HH 135 Enumerated HH 18	Backward village Rainfed agriculture Absolute landless HH 19.7 percent
Piedmont Plain (Northern and Eastern Piedmont Plain)	Biraidakuni/Haluaghat Union/Haluaghat Thana/ Mymensingh	Total HH 177 Enumerated HH 18	Flash flood Garo/Mandi people Absolute landless HH 10.6 percent

Source: Field Survey, 1995–1996

Table 6.2 Number of households facing scarcity in the different (Bangla) months

Months	Active	Meander	Estuarine	Deltaic	Tidal	Coastal	Basin	Piedmont	All	Rank
Baishak-Jaistho	2	7	–	2	–	2	–	5	18	6
Ashar-Srabon	11	–	19	1	2	1	10	4	48	4
Bhadro	11	31	5	5	–	–	25	–	77	3
Ashwin	9	43	21	7	1	2	26	15	124	1
Kartik	4	42	21	3	1	2	15	10	98	2
Agrahayon	–	–	–	–	–	–	–	–	0	–
Poush	–	–	–	1	–	–	–	2	3	8
Magh	–	–	–	1	1	–	–	–	2	9
Falgun	2	4	–	4	3	–	2	–	15	7
Chaitro	1	12	–	7	4	1	6	6	37	5

Note: *Baishak* is the first month in Bengali calendar (it starts in mid-April) and *Chaitro* is the last month (it ends in mid-December)

location, Bhadro, Ashwin and Kartik (mid-August and mid-November) are the three months of scarcity/deficit followed by the months of Ashar, Shrabon (mid-Juneto mid-August) and Chaitro (mid-March to mid-April); the months of Baishakh and Jaistho (mid-April to mid-June) have also been identified as months of stress by many while still other experienced hardship in the month of Falgun (mid-February to mid-March). The minimum period of scarcity experienced was apparently two continuous months while the maximum revealed has been for a period of six months up to seven months for some households in the Piedmont and Basin regions. Evidently households that faced a period of six month deficits, depending on location experienced it in the months of Ashwin, Kartik, Baishakh, Jaistho and Ashar, Shrabon or in the months of Bhadro, Ashwin, Falgun, Chaitra and Ashar, Shrabon. This pattern of a lean season coincides with the documented lean season (Habibullah 1962; Rahman 1996), although for the study households, the months of scarcity are seen as longer and more spread out over the year than the general conditions depicted in the above sections. Duration of scarcity apparently depended on the location of households.

3.2 Spatial variation of scarcity

The BIDS (Rahman 1996) had identified a regional pattern of lean period in terms of employment and income but the variations noted were according to the broad administrative category of divisions and greater districts of Bangladesh. This study makes it apparent that households encountering deficit months not only vary spatially, as shown over such broad administrative divisions, but variations are encountered over a much smaller space. Just as there exist differences in the physical characteristics of terrain and soil, so their climatic variation of temperature and rainfall controls agriculture and influences greatly the employment opportunities and income of the rural people. It is apparent that there exists a great variation in the proportion of households experiencing seasonal scarcity over the months. Variations also exist in the proportion of households facing scarcity among the different regions within the floodplain. When all households of the Basin plain report scarcity/deficit during certain times of the year, only about 21 percent of households of the Tidal plains and 14 percent of households in patiya (coastal) report of such condition. On the other hand, the proportion of households facing seasonal deficit in the Active, Meander and the Piedmont plains is quite – it ranges from 66.6 percent (Piedmont) to 71.4 percent (Meander) to 77.4 percent in the Active plains. Variations in seasonal scarcity of households over the plains therefore depend on a variety of factors ranging from over all ecological conditions of the area to individual characteristics of the concerned households.

3.3 Household scarcity through seasons and regions

Monsoon accompanied by flood bring misery to the people of the floodplains. The households in the Basin and Active plains are evidently the hardest hit. Not only do the households have less/no work during the month of the rains (ashar-shrabon) but households located near river are subject to active fluvial action of erosion/deposition more so in these seasons. Moreover, the population of the flood prone areas experience prolonged period of sickness and disease meaning having to undergo loss of income if the earner falls sick and/or erosion of income in case of other members falling ill. Further, for many households a large portion of the household income has to be expended for house repair after the flood water has receded. It is interesting to note that while households in flood prone areas of the Active and Basin plains (also to some extent households of the Estuarine and Meander Plains) relate their stress period to annual flood occurrences during Ashar-Shrabon and Bhadro, households within the range of tropical cyclones (Estuarine and Tidal) do not relate household scarcity to severe storms/cyclones. This could be due to the fact that whereas floods are regular annual features, severe cyclonic storms and hurricanes are not, although it may be mentioned that the months of April/May (Baishak pre-monsoon) and October/November (Kartik post-monsoon) are vulnerable to natural hazards. The months of Falgun-Chaitro and Baishak-Jaistho are the months of general (soil) water deficit – characterized often by drought conditions. The fact that drought coincides with the normal lean season in the agricultural calendar makes it all the more severe. The households of the Tidal, Deltaic, Coastal and the Piedmont plains evidently experience scarcity during this time of the year, with more households of the Tidal plain experiencing hardship than the others. Most households of the tidal plains faced hardship during the months of Falgun-Chaitro because of the saline condition of the soil; salinity, which is greatly reduced during the monsoon season, rises in the dry months and very badly affects crop cultivation and production.

Seasonal household scarcity, as discussed previously, evidently is related to the primary occupations that households are engaged in. Among the occupational groups facing seasonal scarcity, agricultural wage labour households form the majority (60.1 percent), followed by cultivators (16.6 percent) and van-rickshaw pullers (7.07 percent). About 6 percent and 5 percent of the households in deficit belong to petty traders and day labourers respectively. With the bulk of the rural households dependent on agriculture and agricultural wage labour as a source of income, it is only natural that it will be they who face scarcity in a greater proportion compared to others. It is, however, the duration of the deficit period of households and household characteristics that is important. This leads to the variations in the capability of each household to withstand scarcity, influencing in the process the strategies that each household evolve to meet recurrent scarcity and crisis that befall them.

4. The coping strategy of affected households

Landless households, with limited and irregular employment during the different times of the year, apparently devise various ways and means to overcome and cope with the hardship period that arise due to 'household scarcity'. As scarcity is a function of 'needs' and 'wants', of requirements and desires – within the landless/land poor households are those who during scare times cannot meet the basic needs and also those who, being able to meet the basics, need more resources to fulfil other wants. Both categories of households face scarcity and have strategies to overcome this scarcity. To examine the coping strategies to meet household scarcity this study takes into account all needs and wants that may arise in the resource poor households of the floodplain.

5. Nature of loans and household asset dispossession

Evidently, rural households, depending on their geographical location, experience deficit months ranging from two to even six months. This is the time when many households fail to provide the basics – daily food to household members or treatment for the sick. In the resource constrained households, the difficult/deficit months or 'stress' periods can very easily turn into 'crisis' times. Sickness of the household head or any other member always means a strain in the household budget but sickness during the scarce months of work and income puts the entire household in a state of crisis, distress and despair. Households then have to gear all their activities in trying to provide daily food to the members on one hand and medical treatment for the sick on the other. Furthermore, resources are also scarce for those who want to purchase farm inputs to cultivate the land that they had managed to rent or repair the house that had been damaged by the flood/monsoon rains. For some, scarcity is faced while trying to meet costs of a parent's funeral or a daughter's marriage. Loaning from various sources and sale of household assets are strategies that households resort to in times of these needs. Evidently, the act of loaning is very common – more so than selling household assets. The survey revealed that the proportion of households with current loans is much higher (57.8 percent) than the proportion who had resorted to sale of household assets the previous year (18.4 percent). Previous and current year loan and sale data helps to suggest that about more than half of the floodplain households are always coping with household scarcity by resorting to taking loans and selling assets, if any.

However, comparing the proportion of households facing yearly deficit and the proportion that resorted to sale and/or loan (in the previous year), an interesting feature is apparent. Evidently, where the proportion of household loaning was more, the proportion of households selling off assets was less (Estuarine, Tidal and Deltaic). In Piedmont villages, more households sold assets and fewer households borrowed. With about 67 percent of households experiencing a deficit, only 11 percent had loans, but 19 percent had sold items of value in the Piedmont. Again, the largest proportion of households experiencing seasonal scarcity (Basin 100 percent, Active 77.4 percent; Meander 71.4 percent) have loaned and also sold assets more or less in the same proportion.

6. Source of household loan/credit and items of sale

In household borrowings and/or asset dispossession, sources of credit and items of sale are evidently important indicators of household condition. These elements vary with the household condition of the one who loans/ sells. Sources of credit depend on factors like individual linkages within the network of relations and patrons and credit worthiness. Overall, household conditions affect repayment capacity of households – with capacity being dependent on the purpose of, frequency and amount of initial loan taken. The majority of households, (70.05 percent) evidently, tapped private sources comprising of friends, relations, neighbours and village patrons/wealthy persons for credit in times of need (Table 6.3). *Mohajons*or professional village money lenders known for their high rates of interest have been approached by about 8 percent of households. As compared to the informal sources, the formal/ semi-formal source comprising credit institutions like the Grameen Bank (GB), Krishi Bank (KB,) PDOs like the Bangladesh Rural Advancement Committee (BRAC) are not being

Table 6.3 Sources of current loans

Sources	Active	Meander	Estuarine	Deltaic	Tidal	Coastal	Basin	Piedmont	All
Relations/friends/neighbours	9	29	44	16	9	4	15	12	138
Mohajon	2	2	6	–	2	–	2	2	16
NGOs like BRAC/KB/GB	3	5	10	5	2	1	–	3	29
Local social organizations	–	8	–	–	–	1	4	–	12
Others	–	–	–	1	–	–	–	–	2
Total	14	44	60	22	13	6	21	17	197

frequented very often by the households; only about 15 percent of households borrowed from such formal/semi-formal organizations. Sources of current loan for all households all over the floodplain present a similar picture – there does not seem to be any regional variation in this aspect. Depending on the importance and immediateness of the purpose and the amount of money required, the number of credit sources varied.

The items of value that households sell in times of need reflect household condition to a great extent. Apparently, household assets that respondents sold to meet various needs included cattle, goat, poultry, homestead trees, brass pots/pans, furniture, boat/rickshaw, gold/silver ornaments and land. Cattle/goats make up the major share (50 percent) among the items of value for the household to sell, followed by homestead trees (19.6 percent), land (8.9 percent) and gold/silver ornaments (7.1 percent). In other words, household livestock and homestead trees were the main assets that households sold in times of need – together the two items accounted for about 70 percent of the total proportion of items sold.

7. Reasons behind sale of household assets and household loans

Apparently, households have been observed to have sold multiple items for a single reason which indicates that most households had assets that were not of much value – at least not enough to meet the cost of purpose. The reasons for such sales were to utilize the cash money mostly for household consumption (46 percent), followed by treatment of illness (11.1 percent) and marriages of daughters (8 percent). Besides, households sold to pay children's examination fees, repay loans, free mortgage property, buy assets (sell the cow as it is not giving milk to buy a rickshaw), make/repair house, rent in land, purchase farm inputs and also because there was no work after the floods (more so in Sunamganj). None, it may be noted, resorted to sale of any household items of value in the village of Patiya. The reasons for loans/debts, though similar to the purposes of sale of household assets, are varied in nature. This is because the number of households involved in credits transaction is many and a single loan met multiple needs. Apparently, the proportion of households who had loans last year is less than the proportion of current borrowers – this may not mean that the household conditions have deteriorated within a year. A possible explanation of a lower number of households could be due to a memory lapse, as data collected has been based on the recall memory of respondents.

7.1 Household landlessness

Empirical finding reveals that households have parted with landed property for a variety of reasons-not all of them in a desperate situation. Land has been sold wilfully and the reasons behind land loss were beyond control. Land loss here is a result of some external factors and conditions. Land sale and/or land loss have thus been used concurrently, as both are mechanisms that lead to the land poor and the landless state of a rural household. For the floodplain as a whole, 32 percent households under the present head have experienced land dispossession due to a variety of reasons. Apparently there appears to be spatial variations in the proportion of households that have undergone land dispossession. The proportion ranges from about 68 percent in the Active plain to only about 14 percent in the coastal village. For the rest, the proportion of households that have undergone land sale/loss is about 33 percent in the Piedmont, closely followed by households in the Meander plain (32.1 percent). About a little over one-fourth of the households of the Basins, Tidal, Deltaic and Estuarine, have also undergone land dispossession within the lifetime of the respondent. The actual reasons for which households sold or lost their land have been given in Table 6.4. The same (tangible) reasons spanning two generations have been categorized into nine broad groups under the headings of basic needs, social obligations, investments, economic compulsions, legal situations, political circumstances, ecological conditions, household characteristics and personal traits and behaviour. It presents the various reasons recorded for the different study areas over the floodplain. Based on apparent importance, the following paragraphs discuss a few of the reasons for landlessness (chosen from among the broad categories).

7.2 Spatial variations (of landlessness)

Landlessness, if analyzed separately for various locations within the floodplains, make it apparent that not all regions have experienced landlessness due to the same set of reasons (Table 6.4) and neither have similar reasons been caused due to same set of factors. In other words, there is a variation in causal relationship and connections of factors and reasons of landlessness. The importance of *consumption/food* needs as a cause of rural landlessness may be perceived from the fact that this is the only reason for which households in all three different locations have undergone a certain degree of land dispossession. The households of the Basins, Meander, Estuarine, Deltaic, and the Piedmont plain have been most affected by consumption needs. The spatial variation in the

Table 6.4 Actual reasons for household landlessness over the floodplain

Reasons*	Active	Meander	Estuarine	Deltaic	Tidal	Coastal	Basin	Piedmont	All	Rank
Basic needs	7	56	46	12	20	3	35	33	213	1
Social obligations	1	10	4	3	13	11	12	11	65	4
Ecological conditions	83	5	22	2	–	–	8	3	123	2
Investments (on/off farm)	1	10	2	1	2	–	3	2	21	7
Economic compulsions	–	8	–	2	6	–	3	7	26	6
Legal situations/law and order	1	10	14	18	10	1	4	9	67	3
Political circumstances	–	4	–	–	7	–	1	15	27	5
Household characteristics	–	3	–	–	–	–	4	6	13	8
Personal traits/behaviour	–	8	2	5	4	–	1	1	21	7

Note:*Basic needs include Food, Medicine, Housing; Social obligations include Marriage, Education, Funeral Expense; Ecological conditions include Erosion, Floods, Poor Yield (infertile soil), Low Work Scope; Investments include Buy Land, Buy Agri Inputs, Invest in Trade, Buy Assets, More Profit, Go Abroad/Dhaka; Economic compulsions include Repay Loan, Free Mortgaged Property, Compensate Loss; Legal situations include Litigation, Defrauded/eviction, Disputes/murder, Mortgage Land Lost, Auctioned/nilam, Theft/Dacoity, Vested Property; Political circumstances include Migration/resettlement, Communal Riots, Liberation War; Household characteristics include Death of Head, Old Age, Small Holding; and Personal traits include Alcohol/gambling, Lazy/extravagant proportion of households experiencing landlessness due to household consumption needs evidently has been brought about by ecological conditions of the location aided by socio-economic characteristics of the household and the infrastructural condition of the area. Loss of land to *river erosion* is observed mostly for the villages in the Active Plain. Here it is the single most important cause with about 74 percent of the households having experienced land loss to river erosion. Erosion as a cause of household land loss has also been experienced in the Meander and the Estuarine Plains. *Ill health* of rural people is a major constraint to overall economic condition. A sick person in the household is a liability and hindrance to well-being. Households of the Estuarine and also the Meander have had to sell land when households were burdened with having to bear treatment costs for various illnesses that afflict regularly household members. *Marriage* as a cause of household land dispossession appears to be very important and has been encountered in all regions except the Active and the Tidal Plains. Instances of landlessness due to *legal causes* have been experienced by households in varying degrees in all regions except the Active and Coastal Plains (Patiya). However, being *defrauded/cheated/forcefully evicted*, are acts quite common in the Tidal Plain. Here, landlessness due to flaws in land administration systems and exploitative rural environments accounts for more than one-third of all reasons and it ranks the highest among all causes. Eviction of legitimate owners has taken place when others succeeded in taking over possession by submitting false documents (*benami dokhol*) in collaboration with near relatives and men of authority. Apparently land registration in the Tidal Plain have taken place after murdering the lawful owner; even landless farmers with legitimate allotments to *khas* lands have had to lose possession to local elites. Influential elites also succeeded in taking over Hindu property treated as vested property after registering under false names. Households thus lost control of their landed property when there was either eviction, forced occupation and/or when heads got entangled in costly and time-consuming *litigation* that usually ensue. Reasons of landlessness due to *socio-political conditions* of the country appear to have a spatial aspect but the reasons are more due to the socio-cultural characteristics of the resident population. The reasons thus have more of a social rather than spatial connotation.

7.3 Temporal variations (of landlessness)

It is apparent that there are not many variations in the causes of land dispossession between generations. Apparently, most of the land was dispossessed for basic needs and social obligations more or less in the same proportion in both the time periods (45 percent in the first generation and 50 percent in the second generation). The second generation households had resorted to land sale and experienced land loss more than the present households which is probably because they have been exposed to the forces of landlessness for a longer period of time having completed the household cycle. Also, with more land (compared to the present households) the second generation heads disposed of it whenever a need arose and did not look for alternatives. Mannan (1977) and Ahmad (1978) had also identified such habits as a cause of rural landlessness when respondents blamed previous generations for being reckless, idle and giving in to bad habits. However, it is also a fact that during the past generation, households were large and there were (very) few alternatives and opportunities – fewer than what there is today. Moreover, more second generation households have gone through periods of political turmoil and instability in the country. A comparatively large number have been implicated/entangled in legal complications, probably due to the great social upheaval and turmoil of the time. Conditions such as these led to disruptions of work, activity, an erosion of income and hardships for households. It was revealed during the course of the survey that more Estuarine households had lost land to erosion of the Meghna River in 1950–53 and 1973. The Meghna erosion of the 1950sleft many households in Ramgati landless. This was the time when buffaloes perished in large numbers. Tidal households of Pakigacha reported land loss due to defrauding of second generation households in the decades of the 1940s and 1950s. The Basin household, on the other hand, reported 'bad' years in 1962, 1963 and 1965 – here most land sale took place from the 1960s onwards to the mid-1980s. Although there seems to be a further decline in the dispossession of land in the 1990s in general the trend cannot be taken as conclusive, for it may be mentioned that data related to the 1990s is limited to April 1996. The country in the meantime has experienced another severe flooding in 1998 and a pest attack in the northern districts during the post-flood period. There is a consistency in the persistence of the factors of erosion, consumption and marriage as reasons behind household landlessness over the decades, while no households have reported having to lose land due to political disturbances in the decade of the 1980s and the 1990s.

8. Determinants of rural landlessness in Bangladesh

Spatial differentials, household characteristics, operational spheres and personal traits and habits of household heads/members evidently influenced rural landlessness. These factors interacted with one another to determine household need, influence capability and control available options/opportunities of rural households that led to land sale or land loss – mechanisms that led to the land poor or landless state of a household. Evidently the realities of the landless were local, diverse and dynamic. Figure 6.1 shows the linkages between the various components that affect landlessness of a household. The importance of *spatial differentials* on the landlessness of people depended directly and indirectly on agriculture- people living close to the land at/near/below subsistence level and cannot be over emphasized. Seasonality of agriculture resulted in scarcity of employment and income and variations in terms of duration and severity of household scarcity/deficit was dependent on the geographical location of the household. Seasonal scarcity affected the food availability, nutritional and health status of the people. Spatial differentials therefore directly influenced household needs. Natural hazards of cyclones/storms, floods, droughts are phenomena that deteriorate household condition and are

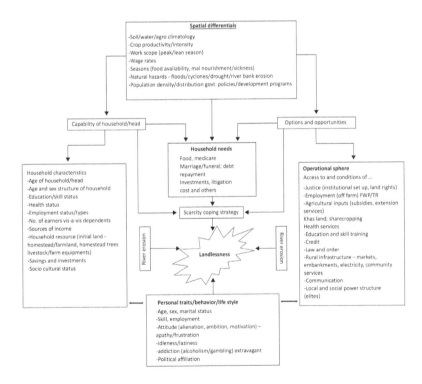

Figure 6.1 Determinants of rural landlessness
Source: Author

totally geographical. *Needs* of rural landless households are similar to needs of landed households but the ability to meet the needs not only differ between the landed and landless but within the landless it varies from household to household. A need for food, medicare, housing, social obligations of marriage and religious rites are faced by all. A need to make investments, repay debts, bear litigation costs, pay for children's education and others apparently varied among landless households depending directly on where the household was located, the state of and relationship with the area of operation and what was the characteristics of the household in terms of its demographic and economic structure and resources. *Households characteristics* in terms of human and other tangible resources determined need and directly affected capability and hence livelihood and landlessness of rural households. Capability to meet household needs depended on demographic, socioeconomic and cultural structure of households. The sphere within which households/heads operated is an important determinant or influencing factor of rural landlessness. The *operational sphere* through its control on opportunities/options directly influenced household needs and coping strategies. Within the operational sphere, local government, police, land administration, social and other institutions are important concerns as they directly shaped the environment in which households functioned.

Personal traits/individual characteristics of household/heads and often household members interacted with other factors and often acted independently to influence landlessness through their attitude and aptitude- employing options, maximizing opportunities, acquiring skills, enhancing capability and influencing needs. Apparently, the dependence and relationship of households/heads with the environment/ecological conditions influenced many of the actions and responses. Landlessness is then a state resulting from an ineffective interaction (by head/household) of various factors operating over time within a particular environment. Household landlessness is a result of a process influenced by and inextricably related to the factors of the general administrative system and the state of politics/government in the country, socioeconomic, demographic, cultural and behavioural characteristics of the people, the phase of infrastructural development of the locality and the agro-ecological conditions of where the household is actually located.

9. Concluding remarks

The study was an attempt at analyzing geographically the determinants of rural landlessness in Bangladesh. The main objective of the study was to capture the dynamics of landlessness as it operated under varied agro-ecological conditions and through causes of household landlessness investigate the spatio-temporal dimension of the phenomenon. The study observes that landlessness cannot be properly explained independent of time and of geographical location. The study has observed that about one-third of the households to have experienced lifetime landlessness and for a large proportion of households landlessness was due to loss of land or to river erosion. The basic issue of landlessness is easy to understand but difficult to solve and where households depend directly or indirectly on agriculture, characterized by low absorption of the rural (additional) labour force, the observed trend can be alarming; unless addressed seriously it runs the risk of jeopardizing measures of our human resources development.

References

Adnan, S. and H. Z. Rahman. 1978. *Peasant Classes and Land Mobility: Structural Reproduction and Change in Rural Bangladesh*. Mimeo.

Adnan, S. and H. Z. Rahman. 1979. *Review of Landlessness in Rural Bangladesh*. Mimeo.

Ahmad, N. 2005. *Slip Trip Tumble: Determining Landlessness in Rural Bangladesh*. Dhaka: The University Press Limited.

Ahmad, R. 1978. *Study of Landlessness in Bangladesh: A Few Cases, Research and Evaluation Division*. Dhaka: IRDP.

Bangladesh Academy for Rural Development (BARD). 1977. *Landless in Rural Society: A Study in Comilla, Bangladesh*. Kotbari Thana: Academy for Rural Development (BARD).

Bangladesh Agricultural Research Council (BARC). 1978. *Incidence of Landlessness and Major Landholding and Cultivation Groups in Rural Bangladesh*. Bangladesh Agricultural Research Council (BARC), Directorate of Agriculture (Extension and Management).

Bangladesh Bureau of Statistics (BBS). 1989. *The Bangladesh Census of Agriculture and Livestock: 1983–84, (Vl), February*. Analytical Report on Census of Agriculture.

Bangladesh Rural Advancement Committee (BRAC). 1986. *The Net-Power Structure in Ten Villages*, Rural Study Series. Dhaka: BRAC (Bangladesh Rural Advancement Committee).

Baqee, A. H. M. A. 1992. *The Settlement Process in Charlands of Bangladesh*. Unpublished PhD Thesis, Department of Geography, University of Dhaka.

Burling, R. 1997. *The Strong Women of Modhupur*. Dhaka: The University Press Limited.

Curry, B. 1979. *Mapping Areas Liable to Famines in Bangladesh*. PhD Thesis, Department of Geography, University of Hawaii, USA.

Faaland, J. and J. R. Parkinson. 1976. *Bangladesh: The Test Case of Development*. London: C. Hurst and Company.

Food and Agriculture Organization (FAO). 1986. *Report of the Expert Consultation on Landlessness: Dynamics, Problems and Policies.* Rome, October 1–4.

GoB. 1998. *The Fifth Five Year Plan 1997–2002.* Planning Commission, Ministry of Planning, GoB (Government of Bangladesh), March.

Habibullah, M. 1962. *Pattern of Agricultural Unemployment.* Dhaka: Bureau of Economics Research, Dhaka University.

Hossain, M. and B. Sen. 1992. *Rural Poverty in Bangladesh: Trends and Determinants.* BIDS Working Paper.

Hye, H. A. 1996. *Below the Line: Rural Poverty in Bangladesh.* Dhaka: The University Press Limited.

Islam, M. A. 1995. *Environment Land Use and Natural Hazards in Bangladesh.* Dhaka: University of Dhaka.

Jannuzi, F. T. and T. P. James. 1980. *The Agrarian Structure of Bangladesh: An Impediment to Development.* Boulder and London: Westview Press, Special Studies on South and South East Asia.

Jansen, E. G. 1987. *Rural Bangladesh: Competition for Scarce Resources.* Dhaka: The University Press Limited.

Karim, A. K. N. 1976. *Changing Society in India, Pakistan and Bangladesh.* Dacca: Nawroze Kitabistan.

Khan, A. R., C. T. Kurien, E. L. H. Lee, S. M. Naseem, R. Nayyar, I. Palmer and I. Rajaraman. 1977. 'Poverty and Inequality in Rural Bangladesh'. In *Poverty and Landlessness in Rural.* Geneva: International Labour Office (ILO).

Mannan, M. A. 1977. *A Survey of Landless and Destitute in Ten Villages of Comilla District.* Comilla: BARD.

Murshid, K. A. S. and A. A. Abdullah. 1986. 'Interdistrict Changes and Variations in Landlessness in Bangladesh'. *The Bangladesh Development Studies,* XIV (3): BIDS; 97–108.

Rahman, A. 1988. 'Development Perspectives for the Most Disadvantaged in Bangladesh'. In I. P. Getubig and A. J. Ledesma (eds.), *Voices From the Culture of Silence: The Most Disadvantaged Groups in Asian Agriculture,* pp. 96–143. Kuala Lumpur, Malaysia: Asian Pacific Development Centre (APDC).

Rahman, H.Z. 1996. '*Mora Kartik*: Seasonal Deficits and the Vulnerability of the Rural Poor'. In H. Z. Rahman and M. Hossain (eds.), *Rethinking Rural Poverty, Bangladesh as a Case Study,* pp. 234–253. Dhaka: The University Press Limited.

Sen, B. 1996. 'Selected Living Standard Indicators'. In H. Z. Rahman and M. Hossain (eds.), *Rethinking Rural Poverty, Bangladesh as a Case Study,* pp. 99–112. Dhaka: The University Press Limited.

Sinha, R. 1984. *Landlessness: A Growing Problem.* FAO Economic and Social Development Series No. 28. Rome: FAO.

Ullah, M. 1996. *Land, Livelihood and Change in Rural Bangladesh.* Dhaka: The University Press Limited.

Van Schendel, W. 1981. *Peasant Mobility: The Odds of Life in Rural Bangladesh.* Assen, The Netherlands: Van Gorcum.

Part II
Physical geography

7 Reconstruction of the palaeogeography and palaeoenvironment of Bangladesh
Geographical perspectives

M. Shahidul Islam

1. Introduction

The natural and built environments which exist today are considerably different from those associated with the prehistoric period. It is quite difficult and challenging to infer the past environment of any location accurately due to the absence of any direct observational records. The dependency on proxy evidences as a tool to unveil the past is questionable. However, at present many specialized techniques are available which can be applied with great confidence to reconstruct the physical and biotic environment of the geological past and the changes in socio-cultural environment associated with prehistoric people. The repetition of similar types of palaeo-environmental investigations can minimize the uncertainties manifold. Scientists are now confidently applying various types of field and laboratory techniques very accurately to reconstruct the past environment of any location.

2. Palaeogeography and palaeoenvironment

The arena of geographical investigation includes understanding the natural processes and how such processes are related to human society and their existence, which are fundamental aspects taken in defining the scopes of geography. Bangladesh has emerged in a deltaic landmass that occupies greater parts of Bengal Basin. The evolution of the basin is the result of an interlink between postglacial land and sea-levels and has been the key question to understand how this landmass becomes so fertile to support large number of population to live. The relative movements of sea-level and the changes of fluvial dynamic processes deriving from decadal to centennial monsoon climate variabilities have given the present geometric shape of the country nearly at about 2000 years

BP. Since then the natural environmental settings and human intervention into nature remain the focal point for geographers to investigate.

3. Palaeo-environmental signatures and techniques to reconstruct

The elements of past environment which can be reconstructed include climate values and pattern, sea-level movements and coastal change, regional and local vegetation pattern, migration routes of plant and animals, soil and water quality and food resources (Tooley 1981). Among those elements some can be reconstructed direct from proxy records, but others require indirect interpretation by available evidences. For example, the maximum transgression limit can be inferred from marine terraces and beach rocks in the field (Monsur and Kamal 1994) and mangrove pollen records in the laboratory (Islam 2001). However, the reconstruction of past climate requires indirect interpretation of proxy evidences. Temperature and rainfall are two major climatic elements that can be inferred from the present or absent of palaesols. The occurrence of drifted sands due to hill slope erosion and their accumulation at lake floor refers to strong palaeo-monsoon (Islam and Debnath 2017).

There are many assumptions, both explicit and implicit, upon which the selection of appropriate techniques and the interpretation of evidences for palaeo-environmental reconstruction are based upon (Tooley 1981). Among those, two assumptions can be considered as the fundamentals of reconstruction. Whatever changes have been occurring in the environment, even at shorter (hourly) time scale, there have been remains of evidences of such changes in nature, either in the sediment layers or in biological (plant and animal) products. The second assumption is the doctrine of 'the present is the key to the past' postulated by, which suggested that the processes operating in the present were the same that had been operating in the past. The application of the Modern Analogue Technique (MAT) through knowledge of present-day ecology makes inferences about the environmental conditions of the past. For example, the analogue of mangrove assemblage of the *Heritiera-Excoecaria* species in present-day Sundabans forest refers to the altitudinal range of past sea-level changes at a 10cm error limit derived from the pollen records of those two mangrove species (Islam 2001).

From a wide range of techniques available to reconstruct the past environment, the initial task is to select an appropriate technique, which largely depends on varying success of biogenic and minerogenic sediment layers. For example, well preservation of mangrove pollen in peat layers of Matamuhuri delta has given the opportunity for pollen analysis of organic sediment to study the evolution of the delta (Islam and Sultana 2011).

4. Lithostratigraphic analysis

The accuracy of palaeo-environmental reconstruction largely depends on the quality of data collected from the field. Poor quality field data will jeopardize the validity of any conclusion and subsequent analysis (Tooley 1981). For these reasons, efforts need to be given on suitable site selection for field data collection and selection of appropriate field techniques. There is a wide variety of earth surface system and depositional environment sediments. Sediment layers, including the ice deposits, are considered to be the natural archives to register the evidence of changes in the earth's environmental system, such as climate changes, vegetation change and sea-level change, human occupancy and culture.

4.1 Ice core

At present, about 10 percent of the earth's surface is covered by permanent ice sheets and the major ice-covered areas are Antarctica, Greenland, Iceland and ice caps of the Himalayan, Alps and polar regions. An ice core sample shows an incremental accumulation of annual layer of snow falls and the lower layer is always older than the upper layer. Within the ice crystals, the palaeo-environmental information, such as atmospheric gases, dust, ash, pollen and radioactive substances of the year of accumulation remained trapped, which can later be untrapped to reconstruct the past climate and past environmental systems, mostly through isotopic analysis. Bangladesh is not a part of the cryospheric environment and scope to explore and study ice core records to unveil the palaeoenvironment of Bangladesh remains remote. However, it is not unlikely that individual geo-scientists, including geographers, from Bangladesh have acquired adequate knowledge and have access to modern laboratory facilities, including isotopic studies of ice core from Himalayan ice caps to unveil a palaeo-geographical account of Bangladesh in the near future.

4.2 Deep sea core

Ocean floors occupy more than 70 percent of the earth's surface and annually 6–11 billion tons of sediments are accumulated in the ocean basin (Bradley 1999). Ocean sediments are composed of both biogenic and terrigenous materials which provide records of past climate change, past sea-level changes and past ocean dynamics. As a littoral country, Bangladesh has ample opportunity to extract and explore the deep sea core records from Bay of Bengal. Deep sea core is an intact sample of sediments collected from the ocean floor by drilling with a long hollow tube. Gravity corer is the widely used technique to collect a deep sea core sample, but it requires a research vessel to move to a desired depth of the sea. Deep sea samples using piston and gravity corers were collected in 1993 by a German vessel RV Sonne (Cruise SO9, and SO126); the study of foraminifera and Oxygen isotope indicates the low rate of sedimentation in the eastern shelf zone of Bangladesh and that the upper continental slope is entirely of Pleistocene sediments. The northern Bay of Bengal, particularly the shelf zone, is one of the largest sediment depo-centres of the world ocean system with total sedimentation of more than 16 km thick. The thickness of Quaternary deposits in the shelf zone varies from 100 to 150 metres of which up to 40 to 700 metres sedimentation is of Holocene period. The palaeo-monsoon signature, the evolution of *Ganges, Brahmaputra, Meghna* (GBM) delta system and Quaternary sea-level movement can be inferred based on core data from the shelf region. The gran size analysis and the ratio of fossil/dead foraminifera frustule give proxies of palaeo-ocean dynamic and sediment influx rate. Oceanography as a systematic science has not yet been well developed in Bangladesh and the progress so far primarily remains confined to biological oceanography, more specifically tomarine fisheries. There remains ample opportunity for the geo-scientists of the country, including the physical geographers, to study deep sea core samples to infer palaeo-ocean dynamics and palaeo-monsoon signals of the sub-continent.

4.3 Lake sediment core

Lakes are widely distributed in all climatic zones. Lacustrine sediments are important archives to register past climate and environmental signals. The alternative thin layers of organic and minerogenic sediment refer to changes in surrounding environmental conditions. The accumulation carbonates in the form of biogenic shell materials have given the opportunities to ease analysis of isotopes of lacustrine carbonate to register paleo-climatic, including palaeo-monsoon, signals. The collection of lacustrine sediment and sediment core are commonly carried out from a floating platform and the coring devices include hand operated gravity and piston corers. Lake sediment records from India, China, Thailand and other Asian countries have been used to reconstruct post-glacial paleo-climate and paleo-monsoon signals in south-central and south-east Asian countries (Cook et al. 2012). They indicated early Holocene intensification of the summer monsoon, which ended with a dry phase at about 6 BP. The central Indian records from lake sediments suggests multiple abrupt monsoon events and climate fluctuation throughout the Holocene. However, Chabangborn and Wohlfarth (2014) argued that strong summer monsoons prevailed over mainland south-east Asia during the Last Glacial Maxima (LGM) in contrast to dry conditions in other parts of the Asian monsoon region.

In Bangladesh, two hilly lakes, the Boga lake and Pukurpara Lake in Bandarban and Rangamati districts, respectfully, are the natural archive to preserve very good quality undisturbed sedimentary records, which can be used to infer regional palaeo-climate changes and generate additional new data for the Asian paleo-monsoon climate model. Despite adequate potentiality for palaeo-environmental reconstruction, these two natural lakes remained unexplored or even unknown to most of the geo-scientists of the country. Islam (2005) for the first time measured that the maximum water depth of Boga lake is 35m and that of Pukurpara lake is 26m (in Debnath and Islam 2015). Islam and Debnath (2017) made the first attempt to collect bottom sediment core using a hand operated piston corer from a locally assemblage floating platform (Plate 7.1). The collection of seven cores from Boga lake and six cores from Pukurpara lake is a breakthrough by Bangladeshi geographers, to explore lake sediment with the view to reconstruct the palaeo-monsoon of the country. Figure 7.1 shows the sequences of thin organic and minerogenic layers in Boga lake and Pukurpara lake, respectively. Such alternative layers of different geochemical properties, organic carbon content and varied grain sizes refer to fluctuations in regional climatic and environmental condition, even for a shorter cycle.

4.4 Lithology

Unlike other sediment sources, lithostratigraphic records from land surface have widely been explored globally for palaeo-environmental reconstruction due to their close connectivity, ease excess, robust utilities, low cost for field work and wider range of environmental coverage. Lithostratigraphic records are mostly obtained from outcrops of hill slopes, excavated profiles and boreholes. An outcrop is the visible exposure of unconsolidated sediment layers of hill slopes and eroded riverbank sides showing their embedding in a series of layers

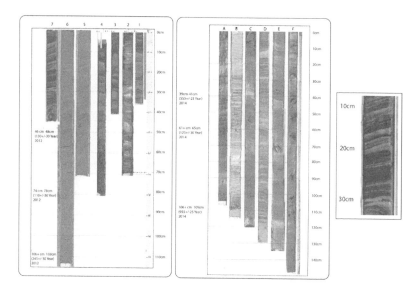

Figure 7.1 The litholo-facies of each core at Boga lake (left) and Pukurpara lake (right). Among the six cores from Pukurpara, two cores (D–E) were collected by Dr. J. M. Garnier (a French geologist) in 2014 and the remaining (A–C, F) were collected by the authors. The last one (F) is the longest core of 150cm long, which has been carried to laboratory analysis. Each core shows a series of embedded very thin lake varve deposits as signals to surrounding environmental pulses, primarily derived from monsoon intensity. The inset is an illustration (zoom) of alternative organic and inorganic layers in core seven at Boga lake, between 10 and 30cm in depth.

Source: Developed by author

of different thickness and grain sizes. Outcrop allows direct observation of sediment types and orientation features for their drifting types, palaeo-current, depositional environment, structural features, bedding plane, fold axes and preservation of fossil evidence. On the other hand, a sediment profile from an excavation, either made to retrieve sedimentary structures, or what was readily available from river/canal bank, pond edge, brick field or any other ongoing or abundant digging sites, clearly shows different layers of sediment of distinctive characters. For clear visibility of each layer, each profile is made vertical, cleaned with a knife and layer boundaries are traced using a matchstick technique (Islam 1996).

It is common practice to retrieve stratigraphic records in the fields by using a number of hand operated samplers. The type most commonly used is the gauge sampler to retrieve an undisturbed core of 1m long with a 2.5cm diameter. This gauge sampler is the most efficient, robust and reliable for a quick and routine survey. A Russian-type sampler is very useful for retrieving a semi-circular half core of peat sediment of 50cm long with a 5cm diameter, without disturbing the stratigraphy. Dacknowski and Augurs of various types are also used in the field. Each sampler is used to collect sediment core from desired depths by joining an additional extension rod of 1 metre long. The selection of appropriate type of sampler largely depends upon the problem under investigation and logistics available.

4.5 Field description of sediments

The first and most important step in lithological studies is the description of sediment in the field. The common practice is to record only textural composition, such as sand, silt and clay, in the field, which might have importance in an engineering investigation, but in palaeoenvironmental studies, such records are not adequate to understand the mode of sediment deposition. Missing records of trace materials could be a great source to retrieve the past. For example, the presence of tiny pieces of charcoal in the sediment layer is evidence of either a palaeo-forest fire or the existence of human occupancy. It is therefore necessary to establish an alternative approach to register both quantitative grain size distribution and also process oriented environmentally induced sedimentation. Troels-Smith (1955) attempted to overcome these difficulties by devising a scheme for recording the physical characteristics of unconsolidated sediment. This scheme is now widely used and has also been accepted by the *International Gorilla Conservation Programme* (IGCP). As this scheme has newly been introduced in Bangladesh (Islam 1996), a brief description is necessary.

In this scheme, there are three elements:

(a) The composition
(b) The degree of humification
(c) The physical properties of a sediment layer can be specified

The composition of a layer is recorded on the basis on a scale of 1 to 4, where 1 indicates 25 percent and 4 indicates 100 percent of the component. The trace amounts of any component in a stratum are represented by the plus (+) sign, where one plus (+) indicates 1 percent of the total components in addition to the major components. A layer may contain one or more components. The main components are: *Turfaherbacea* (Th), *Detrituslignosus* (Dl), *D. herbosus* (Dh), *D. granosus*

(Dg), *Limusdetrituosus* (Ld), *Substantia humosa* (Sh), *Argilla. steatodes* (As), *Aagranosa* (Ag) and *Grana*. The degree of humification indicates the degree of decomposition of the macrofossils that can be observed. It is estimated on a 5points scale where 0 on the scale indicates that the plant structure is unhumified and 4 indicates more or less complete humification. The physical properties of any layer include the degree of darkness (nig. = *nigror*), stratification (strf. = *stratificatio*), elasticity (elas. = *elasticitas*), dryness (sicc. = *siccitas*) and the sharpness of the upper boundary (l.s. = *limes superior*), which are all also estimated on a 5point scale. An example of a complete description for a part of a core at Arial beel is shown in what follows (in Table 7.1).

To avoid confusion and to standardize the drawing and graphical representation, Troels-Smith (1955) also proposed a shade-chart to adapt for each type of unconsolidated sediment (see also Tooley 1981). This scheme has yet remained unnoticed and has not been adopted by most of the geo-scientists of Bangladesh. However, a few attempts so far show that this scheme is very useful, robust and pragmatic in the field to describe unconsolidated sediment of Bangladesh, although the graphical presentation is quite complicated (Islam 1996; Islam et al. 2002; Islam and Sultana 2011; Islam and Islam 2014).

4.6 Laboratory analysis of sediment

Representative sediment samples from the field are carried to the laboratory for appropriate analysis. The most common and easy type of interpretation is the analysis of grain sizes using a standard sieve, hydrometer and/or pippate method (Folk 1974). Grain sizes reflect the mode of transportation and depositional environment of sediments.

Table 7.1 An example of sediment description after the Troels-Smith (1955) Scheme

---

$Ld^2 2$, $Th^2 1$, As1, Dh+
Nig3, sicc1, strf0, elas1, l.s.0
Dark brown to black moderately decomposed peat with roots of herbaceous plants and woody fragments

1st line: 50 percent peat, of which 50 percent is decomposed, 25 percent herbaceous roots, of which 50 percent are decomposed, 25 percent clay and 1 percent herbaceous stems
2nd line: very dark colour, moderately wet, no stratification, layers are non-elastic and the upper boundary is not clear
3rd line: A general description of the layer

Source: Islam, M. S. and Islam, M. 2014.

For example, coarse particles are an indication of high energy depositional environment, such as river beds, natural levee and estuaries, in contrast to clay particles which are deposited in a clam environment, such as the lake floor, lagoon back swamps, peat bogs and sea floor. Statistical techniques, such as frequency curve, mean, median, standard deviation, skewness, kurtosis, bivariant, multivariant, CM patter, factor analysis and cluster analysis are used to determine the mode of sediment transportation and depositional environment in the past. Geochemical properties, such as trace elements, organic carbon, carbonate, bi-carbonate and isotopes of sediments are also indicators of past environmental changes.

5. Pollen analysis

Pollen grain is the male gamete produced within the anthers of flower of an angiosperm plant. This microscopic structure (15–150µ) is protected by a chemically resistant outer layer; the exine remains well buried and preserved in organic sediments, peats, lake muds, ocean floor and even all types of sediment layers for thousands of years. The principle of pollen analysis relies on the assumption that pollen reflects the natural vegetation at the time of their deposition, which provides information about past climatic condition, natural environment, sea-level changes and human occupancy. Since the introduction of the technique by Swedish geologist Lennart von Post in 1916 (in Birks and Birks 1980), it has been widely used to study palaeovegetation, climate and sea-level changes worldwide (Godwin 1940; Jelgersma 1961; Tooley 1974).

The technique of pollen study starts with the isolation of fossil pollen from a sediment sample using standard laboratory procedures (Faegri and Iversen 1989) and the preparation of pollen slides. The major challenge is the identification of species in the slides under microscope. It requires adequate practical knowledge on pollen morphology. The shape of pollen may vary from circular to an elliptical structure and may vary from mono-porate to poly-porate, and mono-colpate to poly-colpate. A standard pollen key (model slides) and reference catalogue can be useful to identify each pollen under microscope. In a tropical country, such as Bangladesh, due to a wide range of plan species it is difficult to develop standard keys and catalogues. However, it is not the textbook, rather it is working experience of an individual scientist and teamwork of mutual sharing that can enhance the better identification and interpretation of fossil pollen. However, despite immense potentiality, the technique has not yet been widely used to infer past-vegetation and environmental history of Bangladesh, particularly during the Quaternary. Alam et al. (1990) attempted to apply pollen fossil to investigate the stratigraphic features of Matamuhuri delta, but the pollen species up to their genera were not identified. Islam (1996) for the first time reconstructed the presence of dense mangrove forest at Panigati (near Khulna city) and at Matuail (near Dhaka city) of Bangladesh by the successful application of pollen technique. Islam and Sultana (2011) have used mangrove pollen records to infer Holocene geomorphic evolution of Matamuhuri delta. Based on pollen records, a few recent attempts have been made to reconstruct the Holocene sea-level changes in Bangladesh (Rashid et al. 2013; Islam and Islam 2014). These studies show that the preservation of pollen grains is very good in all sedimentary sequences, particularly in the peat layers of Bangladesh.

For quantitative representation, at least 150 pollen grains are counted from each layer and pollen are presented in a pollen diagram (see Figure 7.2) The x-axis side of the diagram shows the sediment column and the y-axis shows species identified. Each pollen diagram is divided into number of Local Pollen Assemblage Zones (LPAZ), which are characterized by their more or less uniform composition of flora throughout the zone. TILIA-software is specially designed for quantitative analysis of pollen records and preparation of the pollen diagram. The data presented in the diagram clearly shows the existence of mangrove forest at Arial beel area during the late-Holocene time.

6. Diatom analysis

Diatoms are microscopic unicellular plants, which may vary in size from 5µ to 500µ (Barber and Haworth 1981). Diatoms are found in almost every aquatic environment, ranging from fresh to marine water and also in moist land-surface. Each diatom cell is enclosed by a wall made of silica (hydrate silicon dioxide) and may remain preserved for millions of years in sediment layers. Diatoms are ecologically sensible and have salinity, temperature, water quality, water depth and other environmental tolerances. de Wolf (1982) has classified diatoms into five salinity groups: polyhalobous, mesohalobous, oligohalobous-halophile, oligohalobous-indifferent and halophobous Diatoms are extensively used for palaeo-environmental studies, particularly in palaeo-oceanography. They are useful to trace sea-level changes and have successfully been applied worldwide to reconstruct post-glacial sea-level changes (Tooley 1978; Devoy 1979; Shennan 1986; Ireland 1987).

Like pollen, the technique of diatom study starts with the isolation of diatom frustules from sediment samples using standard method

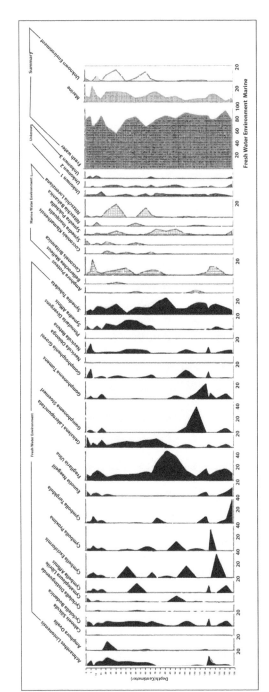

Figure 7.2 Diatom records at Pukurpara lake
Source: Islam, M. S. and Debnath, P. 2017

and preparation of diatom slides. Morphological features, such as size, shapes (centric or pennate), ornamentation of the straie and type of the raphe and frustule symmetry are a few considered to identify each species in the microscope. Diatom as a tool to reconstruct past environment has not yet been widely applied in Bangladesh. Umitsu (1987) identified species in a core at Daulatpur (Khulna) as indication oflithofacies. However, Islam (1996) made the first systematic effort to apply diatom as a tool to reconstruct the Holocene sea-level changes of Bangladesh. The presence of fossils of coscinodiscus and other marine diatom species in sediment layers indicates the existence of shoreline at Matuail (near Dhaka city) during the late-Holocene period (Islam 2001). Some of the identified diatoms are from Matuail. Rashid et al. (2013) reconstructed the evidence of mid-Holocene transgression near Madhupur tract (Dhaka) based on diatom evidences. On the basis of diatom evidence, a few more successful attempts were made to study late-Holocene shoreline movements of Bangladesh. However, the use of diatom evidence of sediment cores collected from Boga lake and Pukurpara lake to reconstruct palaeo-monsoon variabilities and the occurrence of paleo-landslides in the hilly areas of Bangladesh (Islam and Debnath 2017) are the major breakthroughs of palaeo-climate and palaeo-environmental reconstruction of Bangladesh. Local Diatom Assemblage Zones (LDAZ) of the diatom diagram at Pukrpara lake (see Figure 7.2) shows sub-regional micro-scale cycles of intense south-west monsoon.

7. Dating

In palaeo-environmental reconstruction, a chronological framework is an essential requirement. It is necessary to determine the relative or absolute time-line of the events under investigation. Relative chronologies may be build up by correlating the marker horizons in a stratigraphic series. LPAZ may be referred to a regional pollen assemblage zone as a marker to establish relative chronology. For example, without having any results from an organic layer at Arial beel area (near Dhaka), Islam and Islam (2014) projected the existence of dense mangrove forests at the site around 3000 BP, deriving from the relative correlation of marker horizons of stratigraphic series and pollen assemblages at Matuail (near Dhaka) and Panigati (near Khulna) sites (Islam 2001). However, to overcome the uncertainties and ambiguities of relative age dating in palaeo-environmental reconstruction, it is necessary to use an absolute dating technique, although it is an expensive exercise.

From a wide range of radiometric dating techniques, the most commonly used type to determine the absolute age dating of sample is the radiocarbon (^{14}C) dating. This method was developed by Lily and relies on the assumption that the radioactive ^{14}C isotope of carbon is continuously produced in the atmosphere and it enters the living organisms, both plants and animals, being incorporated by them until the organism dies and is buried under the sediments. Once carbon has become fixed in the plant or animal tissue, at death, the radioactive decay of ^{14}C begins at a known rate. The half-life of ^{14}C is 5568 years (Mook and van de Plassche 1986), which makes it useful for dating materials up to at least 45,000 years, but the accuracy diminishes beyond 30,000 years. However, due to the so-called 'Suess effect' (Suess 1955), the practice is to convert the ^{14}C age (e.g., 8210 ±60BP) to calendar age (8199 yrs BCE) to avoid confusion of reconstruction in an age-depth model (Stuiver and Pearson 1993).

Radiocarbon dating is a well-established and powerful technique to provide absolute dating of palaeo index points. It is widely being used to determine sedimentation rate, reconstruction of paleo-vegetation history, post-glacial sea-level changes, development of delta sequences and timeline for prehistoric human occupancy. However, the sample selection for dating is a complicated task due to the risk of contamination of samples either by older organic materials during sediment accumulation (e.g., sediment reworking or hard water effect) or by younger organic materials following deposition (e.g., infiltration and root effect). It is misleading for dating of the drifted materials (wood), which has nothing to predict the timeline of its accumulation in the sedimentary layer due to its over-ageing. It is preferable to use *in situ* organic materials, such as peat, rootlets and plants in their growth position. It requires approximately 2–10gm of organic sediment for conventional ^{14}C dating but for AMS dating only a 5–100 milligram sample is adequate. However, AMS dating is more expensive (about $400–500/sample) than that of conventional dating (about $250–350/sample). There are 142 radiocarbon laboratories worldwide, of which some leading countries are USA (16), China (14), Japan (11), German (9), Russia (7), Italy (5), Poland (5), UK (4) and India (4). Despite immense progress in other scientific areas, Bangladesh has not yet developed any laboratory with the capacity for radiocarbon dating and scientists send their samples to foreign laboratories for analysis. Table 7.2 shows the published ^{14}C data available from Bangladesh. A wood fragment at

Table 7.2 Radiocarbon dates available from Bangladesh

No	Site Name	Depth	Altitude	Lithology	Material	C14 Age (BP)	Source
1.	Daulatpur,	5	-3.00	Peaty clay	Peat	3230±110	Umitsu (1987)
2.	do.	16	-14.00	Peaty silt	Wood	6490±100	do.
3.	do.	13	-11.00	Peaty silt	Wood	6880±130	do.
4.	do	27	-25.00	Clay	Wood	7640±100	do.
5.	do.	34	-32.00	Sandy clay	P. fragment	8890±150	do.
6.	do.	30	-28.00	Clay	Wood	10190±210	do.
7.	do.	43	-41.00	Silty clay	n.a.	12010±210	do.
8.	do.	48	-46.00	Finesand	P. Fragment	12320±240	do.
9.	do.	20	-18.00	Clay	Shell	7060±120	Umitsu (1993)
10.	do.	35	-33.00	Sandy clay	Shell	8910±150	do.
11.	Panigati	3.5	-1.65	Peat	Peat	1210±80BP	Islam (1996)
12.	do	5.75	-3.89	Peat	Peat	5210±60BP	do
13.	do	6.5	-4.64	Peat	Peat	3370±60BP	do
14.	do	8.4	-6.5	–	Wood	5980±60BP	do
15.	do	9.99	-8.09	–	Organic	8210±60BP	do
16.	Kachpur	1.7	n.a.		Peat	3670±60	Islam (1986)
17.	do.	2.65	n.a.		Peat	6060±75	do.
18.	do.	2.85	n.a.		Peat	6460±80	do.
19.	Chandina	1.15	+3.85		Peat	5580±75	do.
20.	do.	2.05	+2.95		Peat	5620±75	do.
21.	Gulshan Lake	n.a.	-2.20		Wood	4040±70	Monsur (1990)
22.	do.	n.a.	-2.80		Wood	4910±75	do.
23.	do.	n.a.	-4.40		Wood	5730±60	do.
24.	do.	n.a.	-6.30		Wood	8940±105	do.
25.	Kalibari	8.90	-4.90		Wood	12780±140	do.
26.	Dakhingaon	2.50	0.00		Wood	4830±75	do.

No.	Location			Material	Age	Reference
27.	Fatulla	3.70	n.a.	Wood	6540±160	Brammer (1996)
28.	Matuail	1.28	+0.60	peaty	2280±80	Islam (1996)
29.	do	1.83	+0.05	Organic	3980±70	do
30.	do	2.19	−0.31	Organic	4080±60	do
31.	do	4.20	−2.32	Organic	6060±60	do
32.	do	5.30	−3.42	Wood	6170±50	do
33.	Sony		+1.8	Peat	1530–1820calBP	Rashid et al (2013)
34.	do		+0.9	Peat	4030–4080calBP	do
35.	do		−1.6	Peat	6410–6670calBP	do
36.	do		−2.8	Lithofacies	6500–7500calBP	do
37.	do		−3.75	Wood	7570–7430calBP	do
38.	Dobadia		+5.7	Peat	5700–5980calBP	do
39.	do		+4.5	Wood	5940–6280calBP	do
40.	Vatpara		+0.4	Peat	5300–5580calBP	do
41.	Chatbari		+3.1	Peat	2760–3080calBP	do
42.	do		+2.1	Peat	4780–4870calBP	do
43.	do		+0.9	Peat	5410–5440calBP	do
44.	Nayanipara		−0.20	Peat	340–3690calBP	do
45.	Bagerhat	3.3	1.3	Peat	2120±70	Goodbred and
46.	do	35	−33	Organic	6940±35	do
47.	do	56.4	−54	Wood	8250±40	do
48.	Barisal	32	−30	Wood	7620±60	do
49.	Khulna	4.3	−2.3	Peat	4770±70	do
50.	Ajmirganj	22.9	−19.9	Organic	5740±50	do
51.	do	39.6	−36	Organic	6320±70	do
52.	do	58	−55	Wood	9390±60	do
53.	Sunamganj	22.9	−19.9	Wood	5480±40	do
54.	do	38.1	−35.1	Organic	5560±90	do
55.	do	73.2	−70.2	Wood	9150±100	do
56.	Netrokona	5.5	1.5	Peat	3810±60	do

112m depth from Jamuna Multipurpose Bridge site (east end) dated 28,320 ±1750 years BP, is the oldest ^{14}C dating age of unconsolidated sediment available in Bangladesh.

8. Paleogeography and palaeoenvironment of Bangladesh

The doctrine of 'Present is the Key to the Past' by James Hutton is based upon the assumption that natural processes, which operate in the earth system now, have always operated in the past. This uniformitarianism concept on slow natural and evolutionary processes has been observed by the earth and natural scientists to give enough time to stream, marine, aeolian, glacial and climatic process to shape up the present landscape and morphology of the earth. The unique topography of Bangladesh includes Chittagong Hill Tract of Tertiary origin, which covers 18 percent of Bangladesh and has been uplifted during the middle-Miocene and Mio-Pliocene orogenies (Khan 1991). The north-south alignment of eastern folded hills, their ranges and large number of anticline and syncline, both symmetrical and asymmetrical, dominate their tectonic origin. However, climate driven hydro-metrological events, along with tectonic and seismic process since the geological past, have had evidential impact on the morphological deformation and reshaping of the landscape of Bangladesh.

8.1 Formation of GBM delta

Deltas are coastal landforms with both subaerial and subaqueous components that are genetically associated with rivers discharging into a standing body of water, such as a lake, estuary, lagoon, sea or the open-ocean shelf. A delta is usually built by a single river, but Ganges-Brahmaputra delta is an exception. Reworking of sediments accumulated at the river mouth by waves, tides and currents should be slow enough to allow delta building to proceed. Whereas waves and tides are key controls of deltaic systems over decades and centuries, delta evolution at longer timescales is most significantly controlled by sea level, climate and tectonic, since the Quaternary Period. A growing knowledge of climate history provides explanation of how high sediment supply, due to increased precipitation and geomorphic landscape disequilibrium, have been playing a role in delta formation even under rapid rates of sea-level rise.

The Ganges and Brahmaputra rivers have formed the world's largest subaerial delta and deep-sea fan system. The rivers are also strongly driven by the summer SW monsoon, when over 80 percent of water and

95 percent of sediments are discharged in the four- to five-month wet season. The immense Ganges river water dispersal system responded to millennial-scale climate change (< 104 years) and that influenced the formation of Ganges-Brahmaputra-Meghna (GBM rivers system) delta system. Some of the major signals of climate change recorded in the delta stratigraphy include thick (20–30 metres) coastal mangrove facies that were deposited during rapid sea-level rise in the early Holocene and a massive volume of sediment trapped during post-glacial transgression.

8.2 Quaternary sea-level change

During the last glacial maxima (LGM), the global sea-level was 121±5 metres below the present (Fairbanks 1989), when most of the shelf zone of Bangladesh was exposed and deeply dissected by a palaeo-river system. Swatch of No Ground was the estuary at the confluence of the Ganges-Brahmaputra river system during the lower stage of sea-level (Chowdhury 1959). During the period of early Holocene deglaciation, global sea-level began to rise rapidly at a rate of 20–34mm/yr between 10000 and 7000 years BP. Additional melt water from Himalayan ice cap and early Holocene strong summer monsoon triggered rapid siltation in the shelf, filling up the incised channels and accelerating the sea-level rise. Umitsu (1987) calculated a high rise, 7.27 mm/yr, of early Holocene sea-level, with a short regression at 12000 years BP along the Bangladeshi coast. The sea-level attained its present level at about 5000 years BP (Benerjee and Sen 1986). Based on lithostratigraphic, biostratigraphic (pollen and diatom) and chronostratigraphic evidences, Islam (1996) made the first comprehensive study of Holocene sea-level changes of Bangladesh. There have been five period of marine transgressions, each followed

Table 7.3 Rates of changes of Holocene sea-level

Transgression	Time period	Maximum rate of sea level change	Average rate of sea level change
Panigati			
Transgression I	9195–6770BP	2.5mm/yr	1.33mm/yr
Transgression II	6615–6415BP	2.5mm/yr	1.29mm/yr
Transgression III	6315–5915BP	3.50mm/yr	2.17mm/yr
Transgression IV	4665–3715BP	1.00mm/yr	0.75mm/yr
Transgression V	2165–1765BP	1.50mm/yr	0.80mm/yr

Source: Islam 2001

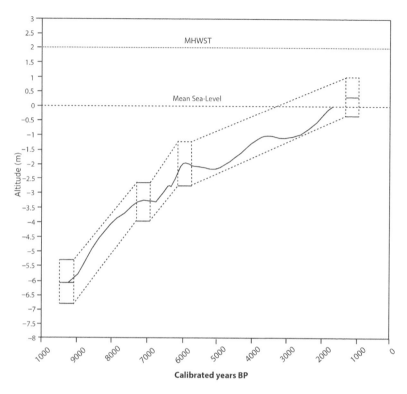

Figure 7.3 Holocene sea-level changes
Source: Islam, M. S. and M. J. Tooley. 1999. 'Coastal and Sea Level Changes During the Holocene in Bangladesh', *Quaternary International*, 55: 61–75

by a regression, and the highest rise was 3.5mm/yr between 6315 and 5915 years. BP (Table 7.3). The sea-level curve (Figure 7.3) presented by Islam and Tooley (1999) shows the Holocene sea-level fluctuations within a band that had never been above the present mean sea level (msl). Rashid et al. (2013), on the other hand, argued for two phases of mid-Holocene transgressions in central Bangladesh, with the highstand of 5m above the present msl at 6000 years BP. Monsur and Kamal (1994) and Monsur and Chowdhury (1996) also argued for mid-Holocene high-stand, although not consistent with their time-period (6000 yrs BP and 5500 yrs BP, respectively). Sea-level scientists in Bangladesh are in agreement on the fluctuating nature of Holocene relative sea-level change, although the rates, directions and timing of such oscillations and their vertical position with respect to present msl are not yet in settlement.

8.3 Palaeo-vegetation

The use of pollen evidence to reconstruct palaeo-vegetation history is a widely accepted technique both in temperate (e.g., Godwin 1940; Devoy 1979) and tropical counties (eg., Mukherjee 1972; Gupta 1981; Barui et al.1986). Evidence available from the Calcutta region shows the existence of early Holocene mangrove in West Bengal (Mukherjee 1972; Sen and Banerjee 1990). Mangrove pollen is well preserved in the organic layer of the coastal and deltaic lithology of Bangladesh. Despite immense scope to infer past-vegetation history, only a few attempts have so far been made and the reconstructions remains confined to three regions: Khulna in the south-west, Chittagong in the south-east and Dhaka in the middle. Islam (2001) projected five phases of mangrove emergence in the south-west coastal belt since 9000 years BP. During phase one (9000–8000 years BP), mangrove forest existed in most parts of Jessore, Narail and parts of Jhenaidah and Magura (Islam 1996). Between 8000 and 7000 years BP (phase two), the sea level was in retreating stage and mangrove forests migrated southward, covering the southern part of Jessore and northern Satkhira. Between 7000 and 5000 years BP (phase three), there were two short periods of marine transgressions (Islam and Tooley 1999) and mangrove forest existed in Gopalganj, Madaripur, northern Khulna and Satkhira region. During phase four (between 5000 and 3000 years BP), mangrove ecosystems existed in the southern part of Satkhira, Khulna, Pirujour and Patuakhali. The northern limit of the present day Sundarbans was established at about 3000 years BP (phase five). Based on pollen and macro fossil (submerged tree trunk) evidences, Islam (1996) suggested the existence of dense mangrove forest at Matuail (south-east of Dhaka city) between 5000 and 4000 years BP. Rashid et al. (2013) also proposed the existence of mangrove vegetation in the north of Dhaka at about 5500 years BP. There was the existence of dense mangrove forest at Arial beel area (south of Dhaka city) only at about 3000 years BP (Islam and Debnath, 2017). Islam and Islam (2014) show the evidence of rapid late-Holocene shoreline migration until the present alignment was obtained in less than a thousand years. They have proposed the model of the Holocene vegetation history with associated environmental change in central Bangladesh (Figure 7.4).

Islam and Sultana (2011) identified the *Excoecaria Agallocha* (Gengwa), a core mangrove species, from the peat layer of Matamuhuri

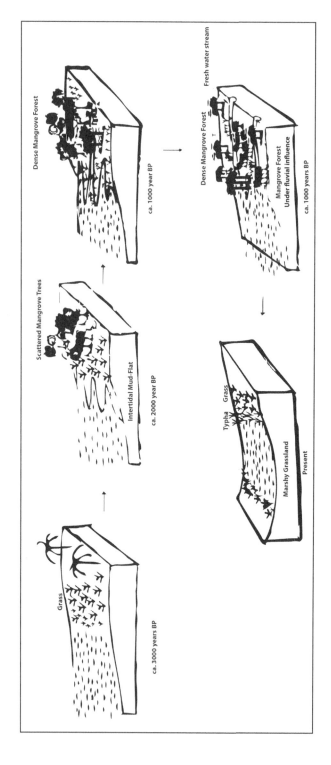

Figure 7.4 A model showing a gradual vegetation succession from beach environment to inter-tidal zone followed by mangrove forest and, at last, marshy grassland

Source: Developed by author

delta. Islam and Sultana (2011) in a comparative study between Pekua (mainland) and Kutubdia (off-shore Island) identified mangrove pollen from peat layers in both the sites and suggested the existence of late-Holocene mangrove forest in the east-coast mainland and also in off-shore islands. The existence of Holocene mangrove vegetation and their disappearance has been the controlling effect of marine transgression and regressions episodes, although their spatial correlations remain unknown. At this stage it is difficult to postulate whether the existence of mangrove forests in south-west, central and eastern Bangladesh are of regional phenomena or emerged as an isolated ecosystem and thus need further investigation.

8.4 Palaeo-monsoon

Palaeo-monsoon refers to the characteristics of monsoon during the geological past. At present, four major sub-systems of Asian monsoon are; North-East Asian Monsoon (NEAM), South-East Asian Monsoon (SEAM), North-West Asian Monsoon (NWAM) and South-West Asian Monsoon (SWAM). Bangladesh located as it is at the tipping point of this entire major Asian monsoon sub-system means that its ecology, environment, economy and society have always been dependent on monsoon characteristics. There has always been a debate regarding the origin of monsoon. Sun and Wang (2005) recognized that the climate system around the Oligicene/Miocene boundary provides evidence of the establishment of the Asian monsoon and, since the Neogene (23 million years), the region has witnessed significant variations in monsoon system. During the Quaternary Period, global climate has oscillated between glacial and interglacial conditions and the Indian monsoon has responded to these climatic changes (Agrawal et al. 2012). Based on lake sediment at Gujarat, India, Prasad et al. (2014) multi-proxy records reveal five phases of paleoclimate changes in Indian monsoon during the Holocene. Between 7500 and 5560 years BP (phase i), the monsoon was intensified, which gradually shifted into a dry phase between 5560 and 4250 years BP (phase ii). After 5000 years BP, the dry condition spread over the greater part of the sub-continent and at around 4200 years BP the condition was excessively dry. Phase iii shows the gradual strengthen of the south-west monsoon after 3500 years BP. Phase iv shows a short pulse of dry period between 3234 and 2709 years BP, followed by somewhat similar to present climatic conditions as phase v. Maher and Hu (2006) established cyclical millennial and multi-millennial rainfall changes in the south-east Asian monsoon region, which has a reverse relationship with the south-west monsoon since the early to mid-Holocene. At about 6000 years BP, when south-west Asian monsoon was intensified, it was dry and weak in south-east Asia and at around 5000 years BP reverse was the case. Such antiphase monsoonal characteristics reflect the relationship between sea-surface temperature and solar forcing. Since the early Holocene, Indian monsoon plays an important role in the global and regional hydrological and energy cycles (Xiao et al. 2004). However, there is a little consensus about the timing, intensity and causes of centennial and decadal scale fluctuations of monsoon precipitation. In the context of Bangladesh, Islam and Debnath (2017) have made the first attempt to reconstruct the palaeo-monsoon signatures during the late-Holocene period. Based on lake sediment they have suggested that the intensity of monsoon had never been uniform and during the last 1500 years there were at least 13 events of alternative strong and weak monsoon phases. These monsoon sequences have had tremendous implications for the terrain morphology and fluvial dynamics of the hilly regions of Bangladesh.

8.5 Palaeo-hazards

Two major palaeo-hazards are palaeo-seismic activities and palaeo-landslides. Palaeo-seismic activity means the earthquake activities in the past. In Bangladesh, the active faults tend to be blind because of high sedimentation rate and thick alluvium mask. The Dauki fault and Tripura-Arakan segments of the mega thrust between Indian and Burmese plates comprised major active structures. The vertical sand dykes in 2 metres thick alluvium deposits infer the records of palaeo-liquefaction as caused by ground shaking from palaeo-seismic incidents in the Habigang area (Morino et al. 2011). The crater deposits at Habigang and the deformed sediment layers later spread in the western profile of Lalmai hill are associated with large-scale earthquake incidents between 1260 and 910 BC and 810 and 420 BCE, respectively. Alam et al. (2012) has given a comprehensive review of regional palaeo-seismic events with an illustration of evidences of palaeo-tsunamies in the Bay of Bengal. The oldest record dates back to about 38,000 BCE, the second oldest event dated to 3000–2000 BCE. The historical records indicate earthquake driven tsunami occurring 1710 CE and 1772 CE affecting the coast of Bangladesh and Myanmar. Paleo-seismic records and historical data are in coherence with frequent occurrence of earthquake incidents at regional scale covering the Tripura-Arakan geological segments.

Palaeo-landslide indicates the landslides of the past. The devastating impact of landslides and secondary instability associated with their unconsolidated debris deposits due to weathering and groundwater seepage are significant geological threats. The determination of age of paleo-landslides has always remained as a challenge and can be correlated with the deposition sequences. The correlation between palaeo-landslides and uplift trends are noticed when the hill slope destabilized due to strong earthquake movements (Martino and Schiattarella 2006). The Central Asian region has records of large-scale bedrock displacement from landslides in the past that have blocked valleys and produced landslide dammed lakes (Storm 2010). Higgitt et al. (2014) recorded the evidences of giant palaeo-landslides to block the Jinsha river in China. Mass movements due to landslides are serious geo-environmental hazards in the Himalayan ranges, including India, Tibet, Nepal, Bhutan and Bangladesh. In the context of Bangladesh, the paleo-landslides have had noticeable implications for the creation of natural lakes on top of the hills of the Chittagong Hill Tract region. Due to intensified palaeo-monsoon sequences there have been tremendous modification of terrain morphology and the occurrence of massive landslides in the hilly regions of Bangladesh. Islam and Debnath (2017) have suggested that Boga Lake and Pokurpara Lake on top of Tertiary hills are not volcanic in origin; rather, these have been created due to intensive monsoon rainfall, which triggered massive landslide and led to the closure of the valley mouths of the hills and the formation of hilly lakes. C^{14} dates suggest that the age of Pukurpara Lake is more than 2000 years BP and that of Boga lake is less than 500 years BP.

9. Concluding remarks

Bangladesh, being a major part of Bengal Basin, provides an interesting entity to study palaeo-environmental changes. To unveil the evolutionary history of any region requires a holistic approach. There are some significant physical and cultural aspects to an area, which need to be properly addressed. The Holocene geomorphic evolution of Bangladesh involves the development of floodplain and its morphology. Most of Bangladesh is now a part of the Ganges-Brahmaputra floodplain. Ganges is the earliest river in this region, but knowledge of its changing courses is not consistent. Only speculation can be made on the changing pattern of its course and the development of its floodplain. At about 2000 years BP, the regression of the sea from the region completed. The soft newly deposited sediments were easily erodible and had made routes for the Ganges to migrate further east. The freshwater of the Ganges carried sediments, which have been deposited on the levee and back swamp of adjacent land.

Peat layers preserve charcoal evidence over geological time. Peat extracted from floodplains can be utilized to reconstruct the history of palaeo-human settlement of this region. Two types of Gramineae pollen, Gramineae large and Gramineae small, are abundant in the peat layers of Bangladesh. Sometimes, the large size pollen of the Gramineae plant species makes for misleading results when compared with pollen species of cereal plants. Such pollen might have scattered from various cereal plants and wild rice. The frequency of cereal pollen is higher in the peat deposits. Charcoal or paleolithic fossil wood are the signature directly associated with human settlement. These microfossils have largely been used to infer the history of human occupancy and their activities during the geologic past (Tooley 1978; Islam 2001). Charcoal has not been extracted from the peat layers. However, charcoal and pollen record suggests the late-Holocene existence of human occupancy and settlement in Bangladesh (Islam 2003). It was then a primitive society and their habitat was within or near the mangrove forest. However, this is only an assumption based on available evidences from very small areas. The archaeologist and the geo-scientist can take the opportunity to infer the palaeo-human occupancy in Bangladesh though a detailed investigation of peat accumulation during the Holocene.

References

Agrawal, S., P. Sanyal, A. Sarkar, M. K. Jaiswal and K. Dutta. 2012. 'Variability of Indian Monsoonal Rainfall Over the Past 100 ka and Its Implication for C3–C4 Vegetational Change'. *Quaternary Research*, 77: 159–170.

Alam, E., D. Dominey-Howes and J. Goff. 2012. 'Tsunamis of the Northeast Indian Ocean With a Particular Focus on the Bay of Bengal Region: A Synthesis and Review'. *Earth-Science Reviews*, 114: 175–193.

Alam, M. K., A. K. M. S. Hasan, M. R. Khan and J. W. Whitney. 1990. *Geological Map of Bangladesh*. Geological Survey of Bangladesh.

Banerjee, M. and P. K. Sen. 1986. 'Late Holocene Organic Remains From Calcutta Peat'. *Bulletin (Geological, Mining, and Metallurgical Society of India)*, 54: 272–284

Banerjee, M. and P. K. Sen. 1990. 'Late Holocene Organic Remains From Calcutta Peat'. *Bulletin (Geological, Mining, and Metallurgical Society of India)*, 54: 272–284.

Barber, H. G. and E. Y. Haworth. 1981. 'A Guide to the Morphology of the Diatom Frustule'. *Freshwater Biological Association Scientific*, 44: 112.

Barui, N. C., S. Chanda and K. Bhattacharya. 1986. 'Late Quaternary Vegetational History of the Bengal Basin. Proc. XI Colloquium on Mlcrpaleontology and Stratigraphy'. *Bulletin (Geological, Mining, and Metallurgical Society of India)*, 54: 197–201.

Bradley, R. S. 1999. *Paleoclimatology: Reconstructing Climates of the Quaternary*. San Diego: Academic Press.

Brammer, H. 1996. *The Geography of the Soils of Bangladesh*. Dhaka: University Press Limited.

Chabangborn, A. and B. Wohlfarth. 2014. Climate Over Mainland Southeast Asia 10.5–5 ka. *Journal of Quaternary Science*, 29(5): 445–454.

Chowdhury, M. I. 1959. *Morphology of the Bengal Basin*. Unpublished M.Sc. Thesis, University of Dacca.

Cook, G. T. and J. van der Plicht. 2012. 'Radiocarbon Dating: Conventional Method'. In S. A. Elias(ed.), *The Encyclopedia of Quaternary Science*, pp. 2900–2911. Netherlands: Elsevier.

Debnath, P. and M. S. Islam. 2015. 'Characteristics of Lake Environment in Hilly District of Bangladesh: A Comparative Study Between Boga Lake and Pukurpara Lake (Bengali)'. *Bhugul-o-Poribesh Journal*, 9: 30–52.

Devoy, R. J. N. 1979. 'Flandrian Sea Level Changes and Vegetational History of the Lower During the Holocene in the Bengal Basin, India'. *The Review of Palaeobotany and Palynology*, 65: 25–35.

de Wolf, H. 1982. 'Method of Coding of Ecological Data From Diatoms for Computer Utilization'. *Mededelingen – Rijks Geologische Dienst*, 36: 95–98.

Faegri, K. and J. Iversen. 1989. *Text Book of Pollen Analysis*. London: John Wiley and Sons.

Fairbanks, R. G. 1989. 'A 17,000 Years Glacio-Eustatic Sea Level Records: Influence of Glacial Melting Rates on the Younger Dryas Events and Deep-Ocean Circulation'. *Nature*, 342: 637–642.

Folk, R. L. 1974. *Petrology of Sedimentary Rock*. Austin: Hemphill Publishing Company.

Godwin, H. 1940. 'Studies of the Post-Glacial History of British Vegetation III. Fenland Pollen Diagram, IV. Post Glacial Changes of Relative Land-and Sea-Level in the English Fenland'. *Philosophical Transactions of the Royal Society*, B230: 239–303.

Gupta, H. P.1981. 'Palaeoenvironments During Holocene Time in Bengal Basin, India as Reflected by Palynology'. *Palaeobotanist*, 27: 138–159.

Higgitt, D. L., X. Zhang, W. Liu, Q. Tang, X. He and S. Ferrant. 2014. 'Giant Palaeo-Landslide Dammed the Yangtze River'. *Geoscience Letter*, 1(1): 6.

Ireland, S. 1987.'The Holocene Sedimentary History of the Coastal Lagoons of Rio de Janeiro State, Brazil'. In *Sea Level Changes*, pp. 26–66. The Institute of British Geographers Special Publications Series. Oxford: Basil Blackwell.

Islam, M. S. 1996. *Relative Sea-Level Changes in Bangladesh During the Holocene*. Unpublished Ph.D. Thesis, University of St. Andrews, Scotland.

Islam, M. S. 2001. *Sea-Level Changes in Bangladesh: The Last Ten Thousand Years*. Dhaka: Asiatic Society of Bangladesh.

Islam, M. S. 2003. 'Mid-Holocene Human Occupancy in Bangladesh: Pollen and Charcoal Evidence From Khulna Region'. In K. Anupama and H. Achyuthan (eds.), *Late Quaternary Environmental Change: Emerging Issues*. Puducherry: Institut Français de Pondichéry, India.

Islam, M. S. 2005. 'Bogalake er Pare, Suprovat Bangladesh'. *Udbodhoni Songlcha*, 6 March, p. 2.

Islam, M. S. and M. Islam, 2014. 'Geomorphological Changes of Arial Beel in Dhaka District' (in Bengali). *Bhugul-o-Poribesh Journal*, 8: 1–22.

Islam, M. S. and P. Debnath. 2017. 'An Approach to the Origin of Boga Lake and Pukurpara Lake in the Hill Districts of Bangladesh'. *Oriental Geographer* (in press).

Islam, M. S., A. Hoque and M. R. Uzzaman. 2002. 'Quaternary Geomorphic Evolution of the St Martin's Island in Bangladesh'. *Indian Journal of Geography and Environment*, 6(1 and 2): 1–24.

Islam, M. S. and M. Islam. 2014. 'Geomorphological Changes of Arial Beel in Dhaka District (in Bengali)'. *Bhugul-o-Poribesh Journal*, 8: 1–22.

Islam, M. S. and T. Sultana. 2011. 'Holocene Geomorphic Evolution of the Mathamuhuri Delta in Bangladesh'. *Oriental Geographer*, 51: 1–22.

Islam, M. S. and M. J. Tooley. 1999. 'Coastal and Sea Level Changes During the Holocene in Bangladesh'. *Quaternary International*, 55: 61–75.

Jelgersma, S. 1961. *Holocene Sea-Level Changes in the Netherlands.* Ph.D. dissertation, Leiden University.

Khan, F. H. 1991. *Geology of Bangladesh.* Dhaka: The University Press Limited.

Maher, B. A. and M. Hu. 2006. 'A High-Resolution Record of Holocene Rainfall Variations From the Western Chinese Loess Plateau: Antiphase Behaviour of the African/Indian and East Asian Summer Monsoons'. *The Holocene* 16(3): 309–319.

Martino, C. and M. Schiattarella. 2006. 'The Palaeo-Landslides of the Melandro River Basin, Southern Apennines, Italy'. *Geophysical Research Abstracts*, 8: 1607–1609.

Monsur, M. H. 1990. *Stratigraphical and Palaeomagnetical Studies of Some Quaternary Deposits of the Bengal Basin, Bangladesh.* Unpublished D. Sc. Thesis, Vrije University. Brussels, Belgium, p. 241.

Monsur, M. H. and M. M. M. Chowdhuri. 1996. Late Quaternary Climatic Fluctuations and the Depositional History of the Bengal Basin. *Journal of Nepal Geographical Society*, 14: 79–86.

Monsur, M. H. and A. S. M. M. Kamal. 1994. 'Holocene Sea Level Changes Along the Maishali and Cox's Bazar-Teknaf Coast of the Bay of Bengal'. *The Journal of NOAMI*, 11(1): 15–21.

Mook, W. G. and O. van de Plassche. 1986. 'Radiocarbon Dating'. In O. van de Plassche (ed.), *Sea-Level Research: A Manual for the Collection and Evaluation of Data*, pp. 525–560. Norwich: Geo Books.

Morino, M., A. S. M. Maksud Kamal, D. Muslim, R. Md Ekram Ali, Md Ashraful Kamal, Md Zillur Rahman and F. Kaneko. 2011. 'Seismic Event of the Dauki Fault in 16th Century Confirmed by Trench Investigation at Gabrakhari Village, Haluaghat, Mymensingh, Bangladesh'. *Journal of Asian Earth Sciences*, 42(3): 492–498.

Mukherjee, B. B. 1972. 'Pollen Analysis of a Few Quaternary Deposits of Lower Bengal Basin'. In A. K. Ghosh, S. Chanda, T. K. Ghosh, S. K. Baksi and M. Banerjee (eds.), *Proceedings of the Seminar on Palaeopalynology and Indian Stratigraphy*, pp. 357–374. Calcutta: Calcutta University, India.

Prasad, V., A. Farooqui, A. Sharma, B. Phartiyal, S. Chakraborty, S. Bhandari, R. Raj and A. Singh. 2014. 'Mid-Late Holocene Monsoonal Variations From Mainland Gujarat, India: A Multi-Proxy Study for Evaluating Climate Culture Relationship'. *Palaeogeography, Palaeoclimatology, Palaeoecology*, 397: 38–51.

Rashid, T., S. Suzuki, H. Sato, M. H. Monsur and S. K. Saha. 2013. 'Relative Sea-Level Changes During the Holocene in Bangladesh'. *Journal of Asian Earth Science*, 64: 136–150.

Sen, P. K. and M. Banerjee. 1990. 'Palyno-Plankton Stratigraphy and Environmental Changes'. *Series C. VI*, 7: 1–100.

Shennan, I. 1986. 'Flandrian Sea-Level Changes in the Fenland. II. Tendencies of Sea-Level State, Brazil'. In M. J. Tooley and I. Shennan (eds.), *Sea-Level Changes*. Oxford: Blackwell.

Storm, A. 2010. 'Landslide Dams in Central Asia Region'. *Landslides-Journal of the Japanese Landslide Society*, 47(6): 1–16.

Stuiver, M. and G. W. Pearson. 1993. 'High-Precision Bidecadal Calibration of the Radiocarbon Time Suess, H. E. 1955'. *Radiocarbon Concentration in Modern Wood. Science*, 122: 415–417.

Sun, X. and T. P. Wang. 2005. 'How Old Is the Asian Monsoon System? – Palaeobotanical Records From China'. *Palaeogeography, Palaeoclimatology, Palaeoecology*, 222(3–4): 181–222.

Tooley, M. J. 1974. 'Sea-Level Changes During the Last 9000 Years in North-West England'. *The Geographical Journal*, 140: 18–42.

Tooley, M. J. 1978. *Sea-Level Changes: Northwest England During the Flandrian Stage*. Oxford: Clarendon Press.

Tooley, M. J. 1981. 'Methods of Reconstruction'. In I. Simmons and M. J. Tooley (eds.), *The Environment in British Prehistory*, pp. 1–48. London: Duckworth.

Troels-Smith, J. 1955. 'Characteristics of Unconsolidated Sediments'. *Geological Survey of Denmark, Series 4*, 3(1).

Umitsu, M. 1987. 'Late Quaternary Sedimentary Environment and Landform Evolution in the Bengal Lowland'. *Geographical Review of Japan*, B60(2): 164–178.

Umitsu, M. 1993. 'Late Quaternary Sedimentary Environments and Landforms in the Ganges Delta'. *Sedimentary Geology*, 83(3–4): 177–186.

Xiao, J., Q. Xuc, T. Nakamurad, X. Yange, W. Lianga and Y. Inouchif. 2004. 'Holocene Vegetation Variation in the Daihai Lake Region of North-central China: A Direct Indication of the Asian Monsoon Climatic History'. *Quaternary Science Reviews*, 23: 1669–1679.

8 Environmental impact assessment in Bangladesh
Applications of geographical tools and methods

Dara Shamsuddin

1. Introduction

Implementation of development projects is pivotal to making progress in a developing nation like Bangladesh. Building infrastructural facilities, the establishment of new settlements and industrial developments are some key areas where new projects are planned and implemented by different government and non-government agencies. Projects like these, either new or expansions of existing ones, require lands to implement different components of the project and thus bring change in different land parameters, like soil characteristics, drainage conditions, slope and aspects, geomorphology and land configuration, etc. In similar fashion, development projects leave impacts on elements of water, air, local plants and animals, thus causing the breakdown of the environmental and ecosystem integrity of an area. In addition, impacts of development projects on local human population, i.e., the social impacts in areas of their production processes (e.g., agriculture, fishery, livestock rearing, etc.), employment provisions, health and education of their children are also assessed through environmental assessment processes. In this context, the Environment Impact Assessment (EIA) systematically reviews the processes of a development project in order to assess how different activities of project in different phases of implementation leave direct or indirect, short or long term, local or strategic, reversible or irreversible, detrimental or beneficial impacts (Glasson et al. 2003) on elements of land, water, air, biodiversity and human population of an area. This assessment helps to understand the magnitude and importance of impacts so that appropriate measures can be taken to reduce the impacts and far-reaching consequences. The EIA results also play roles in developing an Environmental Management Plan (EMP) by the project proponent to ensure the least possible impact on environmental components. Thus, the EIA plays an important role in maintaining balance and harmony of development with the local environment (Ahmed and Sanchez-Triana 2008).

The development of EIA activity started in Bangladesh mainly after the major floods occurred in the country in 1987 and 1988, which triggered the development of the Flood Action Plan (FAP) to mitigate flood impacts. The FAP produced 26 Thematic Papers and among them 16 outlined (in 1992, under the supervision of Flood Plan Coordination Organization (FPCO), the scope of EIA activities for projects designed for water sector. In 2003, the government approved and then published in 2005 the improved guidelines entitled 'Guidelines for Environmental Assessment of Water Management (Flood Control, Drainage and Irrigation) Projects' under the auspices of the Water Resources Planning Organization (WARPO). Later, the Bangladesh Government adopted several policies, laws and rules, e.g., Environment Policy 1994, Bangladesh Environmental Conservation Act 2000 (amended in 2001), Environment Conservation Rules 1997 and Environmental Court Act 2000, aiming to facilitate activities related to environmental assessment.

2. Environmental assessments and their scopes

The definition of the EIA provides an idea about the limits and scopes that the EIA experts should do through a number of steps while performing an exercise. Drawing sharp boundaries around the scope of EIA exercises is a challenging task since the EIA tries to identify and predict the impact of a wide range of interventions, such as legislative proposals, policies, programmes, projects and operational procedures on the environment and on man's health and well-being, and to interpret and communicate information about the impacts (Munn 1979). The array of EIA exercise-domains also make its definitions abound, where different EIA experts, international professional bodies and financial agencies (e.g., the World Bank, the Asian Development Bank) provide different types of definitions of EIA. For instance, the Department of Environment of the United Kingdom mentioned that 'Environmental Assessments (EA) describes techniques and a process by which information about the environmental effects of a project is collected and taken into account by the planning authority in forming their judgments on whether the development should go ahead'(DoE UK 1989). The World Bank has set up three-point aims/principles in describing the necessity of the EIA and to declare the scope of the exercises. The principles are: (a) improving the quality of life, (b) improving the quality of growth and (c) protecting the quality of the regional and global commons. The World Bank emphasizes identifying environmental issues early in the project cycle, designing environmental improvements into projects and avoiding, mitigating or compensating for adverse impacts. The major thematic areas that the World Bank pays special attention to in this regard are natural habitats, water resources, pest management, indigenous peoples, physical cultural resources, involuntary resettlement and forests. Thus the breadth, depth and type of environmental assessments depend on the nature, scale and potential environmental impacts of the proposed projects. The Ministry of Water Resources (WARPO 2005) of Bangladesh describes EIA as

> Environmental Assessments are both reactive and proactive – they are intended to predict the impacts of proposed project interventions and to ensure that environmental requirements are included in project planning. The aim is to identify all the significant negative impacts of a project and to provide recommendations for their avoidance or mitigation, whilst also providing equivalent recommendations on possible enhancement of environmental benefits.

Several other assessment processes are also in practice, such as Social Impact Assessment (SIA), Strategic Environmental Assessment (SEA) and Cumulative Impact Assessment (CIA) to examine likely impacts of new plans and development interventions. Steps and elements of these assessment processes generally overlap with EIA exercises and sometimes these assessment processes work as a standalone approach. These exercises could be categorized in two ways: on the one hand, EIA and SIA assess and foresee likely actual impact conditions (Barrow 2000) against a baseline environmental and social scenario of physical interventions and, on the other hand, CIA and SEA engage in assessing consequential strategic/perceived/cumulative impact conditions.

Discussions on definitions regarding environmental assessment processes presented previously indicate that identifying scopes of environmental assessments is challenging due to the wide array of aspects these methods intend to accommodate in undertaking exercises. It covers all elements of environment, e.g., land, water, air, biological components and human beings. Ahmed and Sanchez-Triana (2008) presume the meaning and scopes of environmental assessments from benefits perspectives, i.e., what benefits could be earned if environmental assessments are properly done and safeguards are ensured while implementing a development project. They mention that 'good management of the environment and natural resources protects health, reduces vulnerability to natural disasters, improves livelihoods and productivity, spurs economic growth based on natural resources and enhances human well-being'. Table 8.1 indicates how different environmental components help people living in floodplain and delta plain areas of Bangladesh by supporting and promoting production processes of various kinds. Understanding this human nature interaction is pivotal to undertaking proper environmental assessments and it also may give an idea about the breadth and scope of environmental assessments.

3. EIA exercises in Bangladesh: legislative frameworks and regulations

Environmental assessment activities are made mandatory in Bangladesh with the promulgation of the Environment Conservation Act (1995) and Environment Conservation Rules (1997). Some other legislative instruments like the Environment Policy (1994), National Water Policy (1999), National Land Use Policy (2001), Wildlife Conservation and Security Act (1992), Protection of Urban Wetlands and Open Spaces (2000) are a few key legislative instruments that guide the environmental assessment activities in Bangladesh. The Department of Environment (DoE) has been playing a major role in regulating environmental assessments and giving project approvals based on results in this regard. The DoE adopted the *EIA Guideline for Industry Sector (1997)* that puts projects in different categories (Green, Orange A, Orange B, Red, see Table 8.2) based on impacts and declared standards for different environmental parameters. Later sector specific guidelines were developed by the DoE for the cement industry, roads and bridges, projects in the natural gas sector (upstream and downstream) and the textile industry.

However, it is important to note that international financial institutions like the World Bank and the Asian Development Bank played important roles in undertaking EIA exercises before the execution of big projects in Bangladesh. As a result, EIA process took place in Bangladesh well before the enactment of related legislative provisions; the EIA exercise for Chittagong Uria Fertilizer Ltd. in 1984–1985, the EIA conducted for Jamuna multi-purpose bridge project in 1987–1989 and the EIA for Karnafuli Fertilizer Co. Ltd. in 1993–1994 are some examples in this regard. Later, after the introduction of Environmental Conservation Rules (1997), entrepreneurs and project proponents came forward to do EIA and obtain environmental clearance from the Director General (DG) of the DoE. However, challenge still remained in post monitoring of the EIAs approved by the DoE and in most of the cases EIAs are not post audited by the proponent.

Table 8.1 Impacts of projects on Important Environmental Components (IECs) and consequential bearing on local people in flood/deltaic plains of Bangladesh

Environmental components (natural resources)	How land resources play roles in supporting people's lives	Disturbance on environmental components and consequential impacts
Land (including soil)	<ul><li>Small-scale agricultural crop production and fishery (capture and culture)</li><li>Homestead vegetable, fruit production</li><li>Fodder for livestock (goat, cow, sheep) gathered mainly from common property sources</li><li>Wider expanse for poultry rearing opportunities</li><li>Fuel wood collection (including fetching cow dung and other fuel sources)</li><li>Water retention facilities like ponds/lakes/other reservoirs</li><li>Groundwater aquifers based on sub-surface lithology</li><li>Opportunities for common property resource collection</li><li>Communication modalities</li></ul>	<ul><li>Agricultural production loss in the area may contribute to shortage of employment for women as daily wage labourers in different agricultural phases like field preparation and sowing/planting, crop harvesting and post harvesting activities</li><li>Loss of crop production in the region (local area) may cause shortage of supply in the market and thus may contribute in food price hike, putting poor women in food inaccessible conditions</li><li>Homestead garden may be washed away and shortage of food and nutrition</li><li>Poultry rearing cannot be done and thus family nutritional aspects and limited earning opportunities may be jeopardized</li><li>Fuel crisis at household level, thus cooking frequency per day might be dropped and that may compel women and their children (and other family members) to consume stale/decayed food putting family health at disease risks; this condition may result increased health cost, challenge in employment generation opportunities</li><li>Water shortage causes many forms of direct (primary) and indirect (secondary) challenges, written in what follows</li><li>Destruction of common property resources in the area may compel poor households to food insecurity conditions and also push them out of the income generation from these resources</li></ul>

Air (considered as a medium that contains important gases; humidity and temperature)	- Vegetation growth, primary production concerns - Sunlight pass-through medium which may be hampered by cloud cover - Human health concerns due to air pollution - Supply of O_2 in a thriving vegetation condition - High temperature in high moisture conditions - Local air quality instigates occurrence of rainfall and acts as agent to control temperature - Air carries water vapor (humidity) which creates necessary conditions for growth of insects, parasites in high temperature conditions	- Food and income insecurity leading to poor health, increased health expenditure and employment at risk due to illness of physically unfit conditions - CO_2 increase may contribute increased primary production upto a certain limit and also will cause temperature increase, which in turn may cause dying of plants, less yield and lead to food insecurity conditions - High moisture also hinders necessary plant respiration process leading to less primary production - Deforestation and less vegetation conditions in the region may not give opportunities for wind break and dissipate energy and thus strong gusty winds may cause huge infrastructural damage - Change in the humidity condition in the air may not give favourable production conditions for many plant varieties (crop phenology)
Water	- Water consumption opportunities for human health - Water for agriculture (crop, fishery, poultry, livestock) - Crop processing - Washing, bathing, cleaning including personal hygiene - Soil moisture, soil fertility - Water pollution; availability of adequate water concerns - Drainage issues including waterlogging. - Support ecosystems, habitats and species - For cooking and food preparation activities - Groundwater for long-term water security - Access difficulty of water	- Less water consumption may put women's health at risk; heath costs increase, loss of employment and income - Loss of agricultural production leading to food insecurity conditions, income generation - Crop processing (post harvesting) hampered; low grade crops may give low market price thus income loss may happen - Water quality deterioration may cause disease, extra health costs - Ecosystem service may be broken down; thus production loss leading to lower income generation - Sense of water insecurity may provide an extra argument (with the set of other arguments) and finally women and their families may take migration decision - Access difficulty puts women and adolescent girls at risk of physical abuse and other forms of harassments. This also puts girls out of school in certain circumstances and women out of work

(*Continued*)

Table 8.1 (Continued)

Environmental components (natural resources)	How land resources play roles in supporting people's lives	Disturbance on environmental components and consequential impacts
Flora and Fauna (plants and animals)	<ul><li>Provide genetic pool (biodiversity), crop cultivars (genetically diverse portfolio of improved crop varieties)</li><li>Ecosystem services (e.g., pollination processes)</li><li>Diversity of crops important for nutritional aspects of food security</li><li>Common property resources (fish, crab, fry collection, fuel, food stuffs) for household nutrition and livelihoods security</li><li>Household level vegetable, fruit production that provides vital nutrients for family members and sometimes provides opportunities to earn money by selling a portion</li><li>Vegetation cover acts as wind break</li></ul>	<ul><li>Impact on flora/fauna may cause production drop; failure of investments (money, energy, time and opportunity costs) might happen; livelihoods security might also at risk</li><li>Crop diversity loss and thus food insecurity (nutritional concerns) might happen; livelihoods security might also at risk</li><li>Challenges in collection of fuel sources (wood, cow dung, etc.) cause time wastes, energy loss and opportunity costs become very high. This challenge is also connected to deforestation, increased fuel costs, less frequent cooking at home, intake of stale food, health risks and related expenditure</li><li>Loss of vegetable and fruits in homestead garden put women and their family members in nutritional insecurity conditions</li><li>Less vegetation cover means less organic matters in the soil and less production potentials leading to food and livelihoods insecurity conditions</li></ul>
Human being (the women as part of environment)	<ul><li>Health of women</li><li>Food security (quantity and quality aspects)</li><li>Water and sanitation</li><li>House and other infrastructure</li><li>Communication and accessibility aspects</li><li>Shortage of commodities/resources may increase competition and related social tension and conflicts</li><li>Skills and capacity aspects to convert opportunities into benefit generation</li></ul>	

Source: Compiled by author

Table 8.2 Procedure of the DoE (Bangladesh) for issuing environmental clearance certificate for different category of industries

Green category projects	Orange A category projects
(a) The application should enclose • General information • Describe raw materials and finished products • A Non Objection Certificate (NOC) from local authority (b) Obtain environmental clearance (c) Renewal every three years	(a) The application should enclose • General information • Describe raw materials and finished products • A Non Objection Certificate (NOC) from local authority • Process flow diagram, layout plan, effluent disposal system (b) Obtain site clearance (c) Apply for environmental clearance and obtain environmental clearance (d) Renewal every year
Orange B category projects	Red category projects
(a) The application should enclose • Feasibility study report • IEE (Initial Environmental Examination) report • EMP • A NOC from local authority • Pollution minimization plan • Outline of relocation plan (b) Obtain site clearance (c) Apply for environmental clearance and obtain environmental clearance (d) Renewal every year	(a) The application should enclose • Feasibility study report • IEE (Initial Environmental Examination) and EIA report • EMP • A NOC from local authority • Pollution minimization plan • Outline of relocation plan (b) Obtain site clearance (c) Apply for environmental clearance and obtain environmental clearance (d) Renewal every year

Source: Compiled by author

4. Case studies of environmental assessments in Bangladesh

4.1 Case study on EIA: establishment of National Oceanographic Research Institute (NORI) in Himchari National Park area

The Government of Bangladesh has established a National Oceanographic Research Institute (NORI) at the Coxs Bazar district located in a hilly terrain dissected by a number of streams and canals. The area is closely connected with costal frontiers characterized by beach morphology. Moreover, the project area falls in the Himchari

National Park with major ecological significance. It suggests that building infrastructures for NORI in the proposed site may require destruction of hillocks and thus create environmental/ecological disturbance in the area. This change in the physical forms of the local landscape may also trigger multidimensional changes in land use pattern, water drainage regimes, air quality and local biodiversity. An EIA was carried out in January 2012 (Islam 2013) to identify the existing pattern of land use and land cover, biodiversity of the site area and the wider region. The field survey showed the factors related to energy and matter flow from the upstream catchments areas to the project site areas. Geographical Information Systems (GIS) along with Global Positioning Systems (GPS) and Satellite Remote Sensing techniques have been used in the survey. These techniques helped to integrate physical layouts of the project components and contour heights with existing land use information (by using overlay techniques). Local government personnel and members of local community have been consulted, by using a Key Informant Interview (KII method) to receive their opinions in regards to the establishment of NORI in the area.

Table 8.3 indicates project impacts on social and environmental components in different implementation phases. Finally, an Environmental Management Plan (EMP) was proposed to reduce and minimize the negative environmental consequences where possible and to bring about necessary design modifications in the layout plan and operational procedures of the institute so that impacts could be kept at the least possible level. Some good practices like tree plantation, construction of guide walls, concrete drainage networks for facilitating water and sediment flow and establishment of ETP are recommended for ensuring safeguards of the environment. But the project components that were identified as factors of significant environmental change/impacts have been relocated in relatively safer places. For instance, the main buildings of the institute were proposed to be overlaid on hillocks. Destruction of these hillocks may contribute in destabilizing the balance of soil mass and slope profile of the locality and also in the wider region. Therefore, the major buildings are proposed to be shifted in the valley areas. It was also proposed that water and sediment flow not be disrupted since any such disruption may cause water congestion and sediment clogging in the adjoining and upstream areas. The physical processes such as accretion and erosion processes when active give the area a dynamic characteristic; also, the man-induced causes bring changes in the area. It suggests that any project intervention may cause further change in the environment and,

Table 8.3 Impacts of project components on different phases of implementation

Project components and factors/activities	Effects (consequential) on environmental and social components
Pre-project (initial) condition	
• Business enterprises will purchase local lands in anticipation of potential tourism business and thus impacting on local land use pattern • Develop agony and anger among local people • Area demarcation by putting physical structures	• Modification of habitats • Change in the land use pattern in the adjoining areas • More commercial activities and increased population density
During project (consequent) construction	
• Construction of buildings • Road infrastructure • Community facilities • Boundary walls • Aquarium • Aquarium tunnel • Jetty/port infrastructure, sea terminals • Wind power generation • Rainwater harvesting facility • Generation of chemical wastes from the laboratories • Withdrawal of groundwater	• Destruction of hills/slope • Imperil cliff structure • Influence the enhanced natural weathering process • Soil erosion • Increased coastal erosion • Gully and sheet erosion • Disruption in the surface water flow • Increased probability of landslide • Increased probability of water and mud flooding • Increased frequency of flooding and bring new areas under flood • Loss of natural biodiversity • Disturbance of sea turtle nesting/hatching grounds • Destruction of (natural) coastal dunes • Sea water pollution from land-based sources • Air pollution • Land use change in the wider locality • Sound pollution • Groundwater recharge may cause dry out local streams • Destruction of visual/scenic beauty • Noise and vibration • Displacement of people and disturbance in production systems

(Continued)

Table 8.3 (Continued)

Project components and factors/activities	Effects (consequential) on environmental and social components
Post project activities and Operation and maintenance <ul><li>Influence the agglomeration of tourism related business enterprises who will establish hotels, restaurants, shops in the area</li><li>Establishment of infrastructure at NORI premise in upcoming three project phases</li><li>More lands might be acquired for area extension</li><li>Waste management issues might be a concern</li><li>Establishment of new transport node/hub</li><li>Water-based recreational facilities development by the private enterprises</li></ul>	<ul><li>Gradual and progressive change in the local land use pattern</li><li>Further disturbance in the physical landscape</li><li>Potentiality of landslide occurrence</li><li>Contamination of land and water from faecal coliform</li><li>More traffic congestion, air and noise pollution</li><li>Coastal erosion</li><li>Impacts from toxic substances</li></ul>

Source: Compiled by author

at the same time, inversely, local physical settings and processes might have significant impacts on the project components.

4.2 Case study on SEA: national adaptation interventions to address climate change induced flood hazards in Bangladesh

A SEA exercise was undertaken (Shamsuddin et al. 2009) to assess the consequences of a number of proposed national level climate change adaptation options in Bangladesh. The study focused on the problems of flood in the floodplain areas of the Brahmaputra-Jamuna and the Ganges-Padma Rivers and came up with certain intervention package proposals. SEA would carry out those interventions, if those are to be implemented in climate change scenario. In carrying out the flood related study, it was found that the process of alluvion/dilluvion is inseparable from flood, and hence was incorporated within the purview of the study. Extensive field work was done in three sites such as Chilmari (Kurigram district); Chouhali (Sirajgonj district) and Zanzeera (Shariatpur district), selected for their representative character. The three sites represented three different physiographic units, two belonging to the Brahmaputra-Jamuna floodplain and the third belonging to the Padma floodplain. However, the main objectives of the field work were to: (a) understand the baseline/existing environmental (both biophysical and social) conditions regarding hazards related to flood and alluvion/diluvion, (b) develop the trends of those conditions, (c) find out local adaptation methods to cope with the hazards from flood and alluvion/diluvion, (d) identify and select suitable local adaptation methods for up-scaling into national level strategies to cope with the hazards predicted under climate change scenarios, and, finally (e) identify gaps in local coping methods and suggest new and effective methods and strategies to fill out the gaps.

The field methods included selection of appropriate sites after several field visits to start with. This was followed by reconnaissance surveys of the selected sites for finalizing appropriate field work strategy and techniques, including developing instruments for a questionnaire survey and Focus Group Discussion (FGD). The detailed field work included a general land use survey using satellite imageries, questionnaire survey, FGD of key informants, observations and interviews and taking photographs as part of field records. Field work was followed by analysis of the findings using both qualitative and quantitative methods, including use of GIS and remote sensing. Brain storming by the team members as well as application of best professional judgment regarding development of climate change scenario and impact assessment played an important role. The results are produced in the form of tables, charts and maps that characterize the existing environmental conditions of the site as well the physiographic units they represent. However, five adaptation packages were selected and SEA performed for all those interventions. These were:

(a) Drainage improvement through basin wise excavation of existing drainage network at community, regional and national levels to allow free flow of flood water, facilitate land raising through sediment deposition and hence aid the delta building process
(b) Water storage through excavation and controlled water level management at community, regional and national levels for dry season and drought period irrigation, fisheries and other uses of water as well as to allow ground recharge
(c) Raising mounds and plinths through excavation to an appropriately higher height at household and community level to address frequent flooding and higher flood water level
(d) Resettlement Support at household and community level for those households and communities displaced due to climate change induced hazards such as erosion, sand deposition, flooding, drainage congestion and the like
(e) Inalienable land rights (in situ) at household level to address the land issues arising out of rapid alluvion/dilluvion as a climate change induced hazard

In order to properly identify the IECs on which impacts are to be assessed, an impact matrix was developed (Table 8.4) that helped to describe the relationship between interventions and the five environmental components, namely air/climate, water, land, biological components and the human components. This helped to identify all the IECs that might be impacted due to interventions. The second table (Table 8.5) was generated and identified, against each IEC, the CC scenario at about 2040, the CC scenario beyond 2040, the scenario without proposed intervention and the CC scenario with proposed intervention. The difference between the two, i.e., CC scenario with proposed intervention and without proposed interventions, is the impact (Figure 8.1). Such impacts were then characterized as being beneficial or adverse, long, medium or short term, and high, medium or low.

The SEA study also extended its scope and carried out a Strategic Cumulative Impact Assessment (SCIA), meaning, the possible cumulative impacts of future policies, plans and programmes on the proposed

Table 8.4 Major environmental components of the study area against adaptation interventions

Climate change adaptation interventions	Environmental components				
	Air/climate	Water	Land	Biological components	Human
Drainage improvement through excavation	Dust during and immediately after excavation	Drainage condition, duration of flood	Siltation, flushing effect on pollution concentration	Riparian vegetation, culture and capture fisheries, migratory bird habitat, insect prevalence	Respiratory health, navigation, access to market
Water storage	Ambient humidity level	Net water storage area in dry season, groundwater availability, groundwater dependence in dry season	Irrigation water availability for agriculture,	Flora and fauna, migratory bird habitats, fresh water habitat, capture and culture fisheries, aquatic biodiversity	Visual aesthetics/scenic beauty, water-based recreation, groundwater cost, rural household water use
Raising mounds and plinth through excavation	Localized dust during and immediately after excavation	Water area increase due to pagar, pond, etc.	Land area, grazing land	Scope for plantation, Avi-fauna habitat, net vegetated area	Scope for institutional infrastructure, protection against flooding, susceptibility to storm/wind to infrastructure, provision for emergency shelter, community common space, common burial/cremation ground, playground

(*Continued*)

Table 8.4 (Continued)

Climate change adaptation interventions	Environmental components				
	Air/climate	Water	Land	Biological components	Human
Resettlement support				Homestead vegetation	Family stability, social integrity, health and sanitation, market access
Land rights			Agricultural productivity, land utilization	Homestead vegetation	Family stability, social integrity and status, health & sanitation, welfare and land rights of future generation, water-based livelihood, Land price, infrastructure, land dispute

Source: Compiled by author

Table 8.5 Strategic level impacts of adaptation interventions on IECs in climate change scenarios

IECs	Existing conditions (2009 CE)	CC scenario (c.2040 CE)	CC scenario beyond c.2040 CE without proposed interventions (Fwo)	CC scenario beyond c.2040 AD with proposed interventions (Fw)	Indicative impact assessment (Fw – Fwo) and impact characterization; (High-H, Medium-M, Low-L; Short term-S, Medium term-M, Long term-L; Adverse-A, Beneficial -B) (5–4)
1	2	3	4	5	6
Drainage improvement through excavation					
Air pollution from dust	Negligible	Negligible	Negligible	Increased	L, S, A
Duration of flood	15 to 90 days	Increased	Increased	Decreased	M, M-L, B
Riparian vegetation	Poor	Worsened	Worsened	Improved	M, L, B
Navigation	Poor	Improved	Improved	Further improved	M, M, B
Drainage congestion	Localized	Widespread	Worsened	Improved	H, L, B
Culture fisheries	Yield moderate	Yield reduced	Yield reduced	Reduced yield loss	M, M, B
Capture fisheries	Yield moderate	Increased	Continued	Decreased yield (more than present)	H, M, B
Access to market	Poor	Worsened	Continued	Improved	M, L, B
Migratory bird habitat	Poor	Increased	Further increased	No Change	–
Net land area (dry season)	Declined – no change	Declined	Continued	Reduced	L, M, A

(Continued)

Table 8.5 (Continued)

IECs	Existing conditions (2009 CE)	CC scenario (c.2040 CE)	CC scenario beyond c.2040 CE without proposed interventions (Fwo)	CC scenario beyond c.2040 AD with proposed interventions (Fw)	Indicative impact assessment (Fw – Fwo) and impact characterization; (High-H, Medium-M, Low-L; Short term-S, Medium term-M, Long term-L; Adverse-A, Beneficial -B) (5–4)
1	2	3	4	5	6
Siltation	Few cm to one m	Increased	Increased	Reduced but controlled	M, L, B
Flushing effect on pollution concentration	Poor	Improved	Continued improvement	Improved	H, L, B
Insect prevalence	Moderate	Increased	Continued	Improved	M, L, B
Water storage during dry season					
Ambient humidity level	Moderate to low	Decreased	Decreased	Increased	M, L, B
Dry season groundwater availability	Declined	Worsened	Worsened	Improved	M, L, B
Freshwater habitat	Poor	Increased	Continued	Improved	M, L, B
Groundwater dependence in dry season	High	Increased	Increased	Decreased	M, L, B
Capture fisheries	Yield moderate	Increased	Continued	Decreased yield (more than present)	H, M, B
Culture fisheries	Yield moderate	Increased	Continued	Decreased yield (more than present)	H, M, B

Aquatic biodiversity	Poor	Increased	Continued	Decreased from CC scenario	M, L, A
Rural household water use	Poor	Improved	Continued	Improved	H, L, B
Net water storage area	Moderate	Declined	Declined	Improved	M, L, B
Irrigation water availability	Limited	Decreased	Decreased	Improved	M, L, B
Groundwater cost (per season)	4,500 to 5,000 taka per hectare	Increased	Increased	Decreased	M, L, B
Visual aesthetics	Moderate	Declined	Declined	Improved	M, L, B
Water-based recreation	Limited	Declined	Declined	Improved	M, L, B
Water edge biodiversity	Poor	Declined	Declined	Improved	M, L, B
Raising mounds					
Localized dust pollution	Negligible	Negligible	Negligible	Increased	L, S, A
Grazing land	Limited	Declined	Declined	Improved	L, L, B
Emergency shelter ground	Limited	Declined	Declined	Improved	M, L, B
Net vegetated area	Limited	Declined	Declined	Improved	M, L, B
Avi-fauna habitat	Limited	Declined	Declined	Improved	M, L, B
Institutional infrastructure	Limited	Declined	Declined	Improved	H, L, B
Protection against flooding	Limited	Declined	Declined	Improved	H, L, B
Susceptibility to storm/wind	Limited	Increased	Increased	Worsened	M, L, A
Community common space	Limited	Declined	Declined	Improved	H, L, B
Common burials/cremation ground	Limited	Declined	Declined	Improved	H, L, B
Playground	Limited	Declined	Declined	Improved	H, L, B
Resettlement support					
Homestead vegetation	Limited	Declined	Declined	Improved	M, L, B
Family stability	Poor	Declined	Declined	Improved	H, L, B
Social integration	Poor	Declined	Declined	Improved	H, L, B

(*Continued*)

Table 8.5 (Continued)

IECs	Existing conditions (2009 CE)	CC scenario (c.2040 CE)	CC scenario beyond c.2040 CE without proposed interventions (Fwo)	CC scenario beyond c.2040 AD with proposed interventions (Fw)	Indicative impact assessment (Fw – Fwo) and impact characterization; (High-H, Medium-M, Low-L; Short term-S, Medium term-M, Long term-L; Adverse-A, Beneficial -B) (5-4)
1	2	3	4	5	6
Health and sanitation	Poor	Declined	Declined	Improved	H, L, B
Market access	Limited	Declined	Declined	Improved	H, L, B
Land rights (in situ)					
Water-based livelihoods	Insecure	Increased insecurity	Increased insecurity	Secured	H, L, B
Land utilization	Insecure	Increased insecurity	Increased insecurity	Secured	H, L, B
Land price	Low	Decreased	Decreased	Improved	M, L, B
Infrastructure	Minimum	Minimum	Minimum	Improved	M, L, B
Land dispute	High	Increased	Increased	Decrease	H, L, B
Agricultural productivity	Moderate	Declined	Declined	Improved	M, L, B
Family stability	Moderate to poor	Declined	Declined	Improved	H, L, B
Social integration	Moderate to poor	Declined	Declined	Improved	H, L, B
Social status of the marginal land owner	Moderate to poor	Declined	Declined	Improved	H, L, B
Health and sanitation	Poor	Declined	Declined	Improve	M, L, B
Welfare and land rights of future generation	Poor and uncertain	Declined	Declined	Improved	M, L, B
Homestead vegetation	Poor	Declined	Declined	Improved	M, M, B

Source: Compiled by author

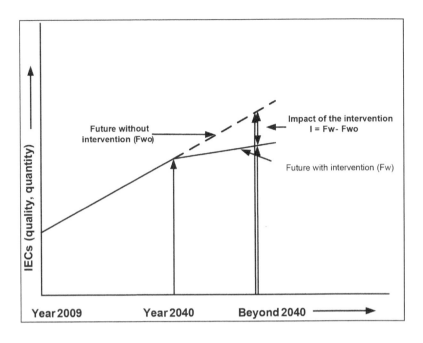
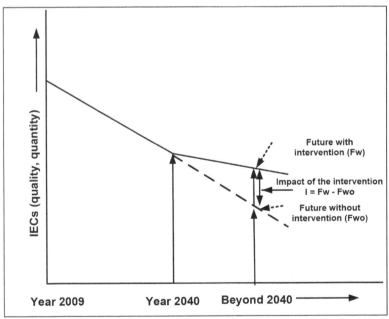

Figure 8.1 Impact assessment of the proposed interventions for two different types of impacts

Source: Islam. M. S. and Islam, M. 2014. 'Geomorphological Changes of Arial Beel in Dhaka District' (in Bengali), *Bhugul-o-Poribesh Journal*, 8: 1–22

intervention package. For individual projects, the CIA is defined as 'the impact on the environment which results from the incremental impact of the action when added to other past, present, and reasonably foreseeable future actions regardless of what agency or person undertakes such other actions' (CEQ 1997). Ideal types of impacts considered in environmental assessments of individual projects are: (a) impacts of environment on proposed interventions, (b) impacts of interventions-on-environment-on-interventions, (c) impacts of other existing projects/interventions, (d) impacts of future projects/interventions, and (e) impacts of projects/interventions outside the border of Bangladesh. The SCIA assessed the impacts against these five scenarios and results showed that the perceived impacts (Table 8.5) would experience further change when other impacts are added with existing effects.

5. Concluding remarks

The results and discussions presented previously suggest that the EIA experts need special kinds of skills by which they could efficiently set the limits of a particular exercise reflecting on the size and nature of the project. The expert also needs comprehensive technical skills so that they can effectively synthesis both qualitative and quantitative data, set the spatial and temporal scales appropriate for the project and type of project location. In concrete terms, expertise on social methods like FGDs, KIIs, Participatory Rural Appraisal (PRA), Case Stories for qualitative data collection, capability on quantitative data collection methods like questionnaire surveys and skills in statistical data handling techniques are important areas of competence required for undertaking high quality environmental assessment exercises. In addition, familiarity with spatial analysis techniques using GIS and Remote Sensing and GPS would help the expert to capture a wide breadth of issues in environmental assessment exercises. Geography students throughout their undergraduate and postgraduate studies learn most of these techniques and have opportunities to apply those in different research works, including environmental impact assessments.

References

Ahmed, K. and E. Sanchez-Triana (eds.). 2008. *Strategic Environmental Assessment for Policies: An Instrument for Good Governance.* Washington, DC: World Bank Publications.

Barrow, C. J. 2000. *Social Impact Assessment: An Introduction.* London: Arnold.

Council on Environmental Quality (CEQ). 1997. *Considering Cumulative Effects Under the National Environmental Policy Act*. Washington, DC.
Department of Environment, UK (DoE). 1989. *Environmental Assessment: A Guide to the Procedures*. London: HMSO.
Glasson, J. and C. Bellanger. 2003. Divergent Practice in a Converging System? The Case of Eia in France and the UK. *Environmental Impact Assessment Review*, 23(5): 605–624.
Islam, S. T. 2013. *Environmental Impact Assessment of National Oceanographic Research Institute (NORI)*. Dhaka, Unpublished Report Conducted for Ministry of Science Information and Communication Technology (MoSICT), Bangladesh.
Munn, R. E. 1979. *Environmental Impact Assessment: Principles and Procedures*. New York: Wiley.
Shamsuddin, D. S., S. T. Islam, M. S. Rashid and M. S. Alam. 2009. *National Adaptation Interventions to Address Climate Change Induced Flood Hazard: A Strategic Environmental Assessment*. Dhaka: Unpublished report conducted for ActionAid Bangladesh.
Therivel, Riki, J. Glassonand and A. Chadwick. 2003. *Introduction to Environmental Impact Assessment*. London: Spon Press.
Water Resources Planning Organization (WARPO). 2005. *Guidelines for Environmental Assessment of Water Management (Flood Control, Drainage and Irrigation) Projects*. Dhaka: Ministry of Water Resources.

9 Climate change crisis in Bangladesh

Sheikh Tawhidul Islam

1. Introduction

Long-term change (about 30 or 40 years) in the climatic variables such as temperature, rainfall, wind pattern, air pressure, sunshine characteristics and related impacts on physical processes like hydro-meteorology, fluvial systems and biological functions are the primary scopes of studying climate change of an area. However, conceptualizations of climate change in Bangladesh are seriously influenced by the United Nations Framework Convention on Climate Change (UNFCCC) and Intergovernmental Panel on Climate Change (IPCC) produced global scale estimates, results and model predictions. These global scale model results may be appropriate to ascertain climate change at a global scale but may not be suitable for understanding local level change happening in the physical processes mentioned previously for different reasons. A few studies like Islam and Neelim 2010; Ahmed and Shamsuddin 2011; Brammer 2014 have been conducted in Bangladesh focusing on climate change investigations but these provide mixed impressions about local level climate change. In the meantime, different policies have been adopted at national level, programmes implemented in different climate hotspots to enhance human resilience and institutional frameworks were evolved and aligned to address climate change impacts. A few other studies at national and local levels were (GED 2012; BANBEIS 2016) also undertaken to understand the climate change impacts more clearly. General Educational Development (GED)'s (2012) study investigated the public sector spending on climate sensitive sectors, while BANBEIS's (2016) study reported the impacts of climate change on educational systems, such as learning competency, impacts on teachers and physical resources of the educational institutions.

It is imperative to mention that the policy planners and other actors working in the fields of climate change in Bangladesh have been trying to develop climate change as a monstrous hazard that may start impacting at a certain cut-off point. In doing that, experts rely on some terminal dates and time (e.g., 2030, 2050, 2070 or 2100) when full blown climate change impacts will be seen under some sets of assumptions. Some argue that impacts of that big monster have already been started. These arguments sideline the main problem and do not allow to ponder climate change as a process and therefore do not provide necessary opportunity to dig deep into the problem. It is also an important point to inquire whether the climate change induced changes are forward progressing (towards a dynamic equilibrium) or changes (interchangeably used as impacts) could be corrected by implementing actions to bring all the changed conditions to the state of what it was before. However, the most alarming thing is that policies and related activities are being adopted without resolving these important questions. Against this backdrop, based on secondary information, this chapter provides geographical explanations of how climate change aspects are conceptualized in Bangladesh, how programme activities are designed, implemented and finally identifies the critical knowledge gaps in areas of climate change.

2. Change in the climate variables in Bangladesh

Climatic disasters (other than technological and environmental disasters) have been the main contexts for Bangladesh communities within which they have been adapting to the challenges for everyday living and gradually strengthen their capacities to thrive. Even the capacities of state agencies have evolved based on the lessons learned from successes and failures of disaster management related actions and process. The Bengal Famine of 1943 (Habib 1995), severe consecutive floods in the years 1954, 1955 and 1956, a severe cyclone in 1970, famine again in the 1980s (1974) due to Brahmaputra flooding and crop failure, widespread and prolonged floods in 1987 and 1988, cyclone stricken in 1991, then cyclone Sidr in 2007 and Aila in 2009 are all known as big triggering events in Bangladesh disaster history of the past century. The major aspect related to climate change with these disasters lies in the fact of whether any change in the climatic variables influence these disasters like floods, droughts or cyclones in terms of degree of intensity, duration, frequency and severity. But research attempts are non-existent in Bangladesh to determine this influence; even research works for determining change in the climate variables are also inadequate. Islam and Neelim (2010), in this regard, show (Table 9.1) that temperature variable for both winter and summer seasons does not show any significant pattern of change (either increase or decrease). Major

Table 9.1 Average maximum and minimum temperature fluctuations of selected weather stations in Bangladesh

Districts	Winter average maximum (average of December and January)				Winter average minimum (average of December and January)			
	Fluctuations	Climate line	Warmest year (deviation from climate line)	Coolest year (deviation from climate line)	Fluctuations	Climate line	Warmest year (deviation from climate line)	Coolest year (deviation from climate line)
Dinajpur ('48–07)	Decrease	24.91	1972 (2.49)	2003 (−2.46)	Increase	11.10	2002 (1.45)	1983 (−1.45)
Sylhet ('56–07)	Increase	25.78	1997 (1.97)	1964 (−1.08)	Increase	13.32	2002 (1.38)	1962, 1978 (−1.27)
Iswardi ('61–07)	No change	25.11	1987 (1.54)	2003 (−1.41)	Increase	11.11	2002 (1.59)	1989 (−1.46)
Dhaka ('53–07)	No change	25.74	1988 (1.56)	2003 (−1.89)	Increase	13.21	2002 (2.09)	1962 (−2.36)
Jessore ('48–07)	No change	26.15	1985 (1.85)	1981 (−1.3)	Increase	11.83	1986 (3.07)	1949 (−1.93)
Hatia ('66–07)	Increase	25.57	2002 (1.03)	1974 (−1.32)	Decrease	15.42	1979 (1.48)	2001 (−1.97)
Coxs Bazar ('48–07)	Increase	27.01	2000 (1.69)	1962 (−1.48)	Increase	15.61	2005 (2.24)	1950 (−1.76)

Districts	Summer average maximum (average of April and May)				Summer average minimum (average of April and May)			
	Fluctuations	Climate Line	Warmest year (deviation from climate line)	Coolest year (deviation from climate line)	Fluctuations	Climate line	Warmest year (deviation from climate line)	Coolest year (deviation from climate line)
Dinajpur ('48–07)	Decrease	33.61	1957 (4.99)	1983 (–4.11)	Slight increase	22.17	1970 (1.53)	1957 (–1.52)
Sylhet ('56–07)	No Change	30.98	1960 (3.27)	1977 (–3.23)	Slight increase	22.14	1949 (1.11)	1986 (–1.89)
Iswardi ('61–07)	Decrease	35.27	1972 (4.13)	1981 (–3.37)	Slight increase	23.56	1999 (1.69)	1976 (–1.86)
Dhaka ('53–07)	Decrease	33.54	1960 (2.41)	1977 (–2.44)	Slight increase	24.14	1999 (1.41)	1977 (–1.49)
Jessore ('48–07)	No Change	35.55	1957 (2.75)	1949 (–3.2)	No change	24.29	1954 (1.26)	1997 (–1.44)
Hatia ('66–07)	Increase	31.91	1979 (1.74)	1970 (–1.86)	Decrease	24.57	1966 (1.53)	1967 (–1.32)
Coxs Bazar ('48–07)	Increase	32.16	2005 (1.74)	1977 (–1.91)	Increase	24.46	1995 (1.34)	1949 (–1.86)

Source: Compiled by Author

observations they suggested are: (a) weather stations in the coastal belts of the country show temperature rise both in average minimum and maximum for summer season. On the other hand, stations located in the northern Bangladesh show a relatively low increase compared to that happening in the coastal belt stations, (b) a significant level of spatial variation has been identified in the change of temperature variable. Islam and Neelim (2010) also commented on rainfall variable; they suggested that (a) no definite evidence was found indicating any significant shift in the pattern of monsoon rainfall, (b) coastal and hill weather stations show high rainfall occurrence compared to the stations located in the floodplain areas and northern parts of the country, (c) there is a general tendency for most of the weather stations to have more rainfall occurrence at the end of monsoon season (i.e., August), rather than equally distributed among monsoon months or happening at the beginning of the season, (d) among the monsoon months (i.e., June, July, August and September), July and August received consistently higher rainfall than the remaining two months, (e) an average of 75–80 percent rainfall takes place during monsoon times in Bangladesh.

Table 9.2 Average monsoon (JJAS) rainfall compared to average annual total; highest and lowest deviation from monsoon average rainfall (Bangladesh annual average is 2300 mm)

Agro-eco zones	Average of monsoon rainfall against average total annual rainfall (in mm)	Highest deviation from monsoon average and year	Lowest deviation from monsoon average and year
Dinajpur ('48–07)	1,452 (1911)	+ 925 (1995)	– 641 (1994)
Sylhet ('56–07)	2,666 (4074)	+ 1,292 (1989)	– 1,299 (1980)
Iswardi ('61–07)	1,120 (1589)	+ 997 (1976)	– 424 (1994)
Dhaka ('53–07)	1,372 (2090)	+ 748 (1984)	– 669 (1958)
Jessore ('48–07)	1,142 (1615)	+ 825 (2004)	– 539 (1958)
Hatia ('66–07)	2,449 (3257)	+ 838 (2001)	– 770 (1989)
Coxs Bazar ('48–07)	2,894 (3661)	+ 1,541 (1987)	– 1,803 (1980)

Source: Compiled by Author

However, in common practice, climate change and disasters (e.g., floods, cyclones and droughts) are used interchangeably in Bangladesh without taking into account their functional connections. It is important to note that no serious attempts have been made to assess the degree of influence of the change in the climatic variable (temperature in this case) on disaster events.

3. Climate change impacts in Bangladesh

3.1 Impacts incurred from sudden big events

The climate change impacts are commonly seen occurring as sudden big events such as cyclones, floods and also as slow and progressive phenomenon like drought conditions, salinity intrusion, sea level rise and related threats. The accounts of impacts caused by sudden big events are briefly given in this section and following section illustrates impacts happening from slow and progressive events. Impact statistics suggest that about 0.63 million people died of natural disasters like cyclones, floods, earthquakes, extreme temperatures and different kinds of epidemic events in Bangladesh since 1950 (EM-DAT 2016). It is important to note that cyclones appeared to be the deadliest disasters in Bangladesh. Cyclones caused about 90 percent (567587) of total deaths of natural disasters that happened from 1950 to 2016. Floods occur annually, reaching their most severe in the months of July and August. The floods of 1988, 1998 and 2004 were catastrophic, resulting in large-scale destruction and loss of lives (51,986 people die of floods from 1950 to 2015). A total of 129 cyclones of different categories hit Bangladesh coasts from 1978 to 2013 with an annual mean occurrence of 3.6. November is the month in post-monsoon season when the highest number of cyclones (31) made landfall, while the month of May received the highest number (20) in the pre monsoon time from 1978 to 2013. The estimated economic loss incurred from disaster impacts in Bangladesh are calculated differently by different agencies. For instance, BBS (2015) in their Impacts of Climate Change on Human Lives (ICCHL) study estimated that the economic loss of disaster impacts was about 2.3 billion USD from 2009 to 2015. On the other hand, the Centre for Research on the Epidemiology of Disasters (www.cred.be, last accessed on 27 October 2018, based in Belgium) estimated 2.9 billion USD from 2005 to 2015.

Lightning has suddenly drawn attention in Bangladesh with its sudden onslaught as something more than a hazard, but rather a disaster. Lightning in 2016 (in the month of April, May and June) caused death

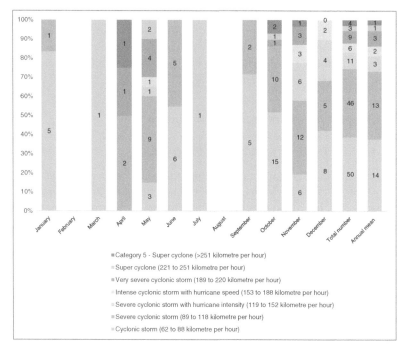

Figure 9.1 Tropical cyclone category
Source: EM-DAT. 2016. *The International Disaster Database*. Centre for Research on the Epidemiology of Disasters –CRED) (www.emdat.be), 2016

of people, loss of property and spread panic across the country. The State Minister, Disaster Management and Relief, Bangladesh informed the Members of the Parliaments on 5th of June 2016 that 142 people had died of thunderstorms this year. This death toll and damage of assets made the Government of Bangladesh include thunderstorm hazards (and associated lightning) as a national disaster. Climate scientists suggest (Diffenbaugh et al. 2013) that in the future the energy available for convection (called Convective Available Potential Energy or CAPE) will increase in climate change induced greenhouse gas forcing conditions. As a high value of CAPE is conducive for severe thunderstorms, we can expect to have more thunderstorms and lightning hazards in the future in Bangladesh. With global warming expected to continue in future, thunderstorms will occur more frequently, given the conducive context for it to form (Samenow 2013).

3.2 Impacts incurred slow and progressive hazards

3.2.1 Drought

Drought is a water stress condition in a time when rainfall is generally expected for plant growth. Incidence of drought in the north-western regions of Bangladesh is quite common where rainfall occurrence is almost half of that occurring in north-eastern regions of the country. June is the month when drought usually occurs as rainfall; June has always had less in north-western parts of the country compared to the other parts of the country. The studies carried out in the 1980s by the Bangladesh Agricultural Research Council (BARC) in order to develop an Agro-ecological Zone (AEZ) database for the country show that monsoon rainfall starts in the first week of June, more precisely from the first of June in Bangladesh, but this starting time deviates 26 days from the first of June in Rajshahi district located in the north-western regions of the country (Brammer 2014). It indicates that the late start of monsoon in drought prone north-western region is not an abnormal phenomenon, which was observed in the 2009 drought that occurred in this region when rainfall was recorded only 146 mm against a 45-year mean of 271mm. It is also important to note that most of the rainfall (105 mm against a total of 146 mm, 72 percent) in that year occurred on the 29th and 30th of June (in 2009). It suggests that understanding drought phenomenon needs more rigorous study using daily rainfall data (not monthly or yearly average) including associated daily temperature records so that PET (Potential Evapotranspiration) rates can be calculated.

3.2.2 Salinity intrusion

One of the immediate and long-term threats of climate change is the ingress of salinity into surface (i.e., river water) and groundwater systems and into the soil. Salinity intrusion in the coastal areas resulted from a number of factors, such as reduction of fresh water flows from Ganges, siltation in the tributaries of the Ganges river systems and siltation in many other rivers due to the impacts of polders. Dasgupta et al. (2014) carried out a study for the World Bank and found that salinity increased from 2ppt to 20ppt at Mongla (south-western coastal regions) in the Passur river from 1962 to 2008.

In addition to surface water salinity, it is of the utmost important to appreciate soil salinity due to its role in agricultural production. SRDI periodically estimates soil salinity for coastal regions of Bangladesh.

The third such report in a row was published in 2010. The study findings suggest that total soil salinity has increased to about 1.056 million hectares from 0.8333 million hectares in about four decades. Currently, out of about 1.689 million hectares of coastal land, 1.056 million hectares (about 63 percent) are affected by soil salinity of various degrees. About 0.328, 0.274, 0.189, 0.161 and 0.101 million hectares of land are affected by very slight (S1; 2–4 dS/m), slight (S2; 4.1–8 dS/m), moderate (S3; 8.1–16 dS/m), strong (S4; >16 dS/m) and very strong salinity (S5), respectively. A comparative study of the salt affected area from 1973 to 2009 showed that about 0.223 million hectares (26.7 percent) of new land is affected by various degrees of salinity over about the last four decades (Table 9.3).

3.2.3 Sea level rise and associated risks

The coastal areas of Bangladesh have been adjusting to Sea Level Rise (SLR) in the Bay of Bengal over the last 11,000 years, at least (Kuehl et al. 2005; Mikhailov and Dotsenko 2007). This happened with the deposition of sediments carried out by three major rivers, i.e., Ganges, Brahmaputra and Meghna, one-third of which is deposited on the active floodplain areas and in the channels of the delta (Goodbred and Kuehl 2000). A small portion of sediment is delivered via waves and tides to the inactive, tidal portion of the delta at a rate about 10mm (highest) per year (Rogers et al. 2013). However, the coastal areas are still subject to adjustments since the average elevation of it ranges from 1metres to 3metres in the south-western and central coastal regions (about 80 percent of the coastal areas) allowing dynamic variables like incoming sediments, SLR and land subsidence to play roles. The elevation of south-eastern coastal zones is about 4 to 7 metres. This low elevation, active delta and dynamic morphology play significant roles in making the region vulnerable to hazards like cyclone induced water surge and the submergence of lands by the high level of sea water. The area is also vulnerable to coastal land erosion and salinity intrusion into the surface and groundwater and soil.

The coastal areas of Bangladesh are generally divided into three categories, e.g., the Ganges Tidal Plain or the Western Coastal Region, the Meghna Deltaic Plain or the Central Coastal Region and the Chittagong Coastal Plain or Eastern Coastal Region. These three coastal regions experience different SLR scenarios; for instance, water level increase is happening at a rate of 5 to 7mm per year in the south-western coastal regions, around 10mm per year in the central coastal regions and about 14 to 23 mm per year in the south-eastern coastal regions (CEGIS 2016).

Table 9.3 Soil salinity in coastal Bangladesh

Region*	Salt affected area (000 hectares)			Soil salinity class and area (000 hectares)												Salinity increase over 4 decades	
				S1 2–4 dS/m			S2 4.1–8 dS/m			S3** 8.1–16 dS/m			S4 >16 dS/m			Area (000 hectares)	percent
	1973	2000	2009	1973	2000	2009	1973	2000	2009	1973	2000	2009	1973	2000	2009		
1	374	417	432	48	93	87	255	119	102	52	161	169	20	48	68	57	15
2	78	78	76	18	24	25	53	27	19	3	19	27	0	7	1	–	–
3	100	106	106	25	16	21	31	33	29	24	46	50	19	72	6	6	6
4	0	26	33	0	10	22	0	5	8	0	1	2	0	0	0	33	100
Bangladesh	833	1020	1056	287	289	238	426	307	274	79	336	351	36	87	101	222	26

*1 – South-western regions (Khulna, Bagerhat, Satkhira)
 2 – Central coastal regions (Laxmipur, Feni, Noakhali)
 3 – South-eastern regions (Chittagong, Cox's Bazar)
 4 – Mature delta regions (Jessore, Narail)
**S3 = S3+S4; S3 = 8.1–12 dS/m; S4 = 12.1–16 dS/m. Survey conducted in May 2009

Source: Soil Resources Development Institute (SRDI). 2010. Saline Soils of Bangladesh. Report prepared as part of SRMAF Project, p. 37

Table 9.4 Financial loss as a result of disaster impacts in Bangladesh

Source	Duration covered	Estimated loss from disaster impacts
Emergency Events Database (EM-DAT)	2005 to 2015	2948 million USD
BBS (2015)	2009 to 2015	2301 million USD

Source: Compiled by Author

Estimating the economic loss of disasters is a challenging task since the consequences of immediate and far-reaching impacts are difficult to track and therefore to calculate in financial terms is challenging. However, a few attempts are made to calculate approximate values where infrastructural and crop damage are generally taken into account. EM-DAT (2016) and BBS (2016) provide estimated figures in this regard.

4. Social dimensions of climate change in Bangladesh

4.1 The nexus between climate change impacts poverty

Sustenance and security of the majority of people, their living arrangements and conditions in Bangladesh have been heavily dependent on the availability, quality of natural resources and also on the access issues to these natural resources. These natural resources, like: (a) land with favourable properties, such as fertility, moisture holding capacity, well drainage systems, (b) biological diversity and resources that provide wide array of choices to cultivate by the farmers, (c) water resources that ensures irrigation facilities, supply important nutrient elements for plants and animals, and (d) climatic conditions like optimum temperature, sufficient occurrence of rainfall, wind direction and flow pattern, necessary sunshine, all in a combined fashion and effect, create an enabling environment suitable for natural resources based primary productions systems in this country.

Any impacts of climate change are mediated through these physical systems/processes, therefore the effect/influence of climate change on these processes are necessary to understand in order to recognize the climate change impacts on food, livelihood and security of people by and large on social systems. However, the physical processes and their resultant natural resources offer different livelihood options to the people and many small livelihoods groups (like small traders,

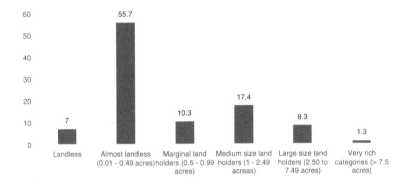

Figure 9.2 Percentage of household by land ownership in Bangladesh
Source: Bangladesh Bureau of Statistics (BBS).2007. *Labour Force Survey 2005–06*.
Dhaka: Ministry of Planning

transportation workers) emerge and perform different activities. These natural resource-based bundles of activities finally compose and characterize primary production systems that includes agriculture, forest and fisheries sub sectors.

The Bangladesh Bureau of Statistics (BBS 2011) suggests that 56.7 million people currently constitute the labour force of the country, where 39.5 million (70 percent) are male and 17.2 million (30 percent) are female. About 25.7 million (47 percent) are directly engaged in agriculture and fisheries sector, of which almost 62.7 percent are functionally poor landless (when the bottom two categories are combined); about 27.7 percent have land resources between 0.5 and 2.49 acres and only 9.6 percent fall in the rich category in terms of holding land resources. This indicates that the people who are engaged in primary production systems by taking the benefits of different hydro-meteoroidal and other physical processes are generally poor. In these areas/sectors some people organize the economic activities and appear as owners or employers, some are engaged as self-employed persons and a major part engage in the form of day labourers (agricultural and non-agricultural), petty traders, crop processors and transportation workers. It is important to note that about 95.40 percent (there are 5.53 million day labourers in Bangladesh when all other sectors are combined, about 25 million people are dependent on them) of day labourers of the country come from the agriculture sector (BBS 2011). This suggests that the occurrence of any impacts, whether it is strong episodes of disasters or slow onset climate change calamities, put these marginal people (who are labelled as hardcore poor, currently comprising 7 percent of whole population) in serious poverty ridden conditions. It happens because they have no land or possess minimum land resources on the one hand and on the other hand land utilization processes remain at multiple and cyclical forms of disaster threats. These conditions create the grounds of poverty climate change nexus.

4.2 Gender dimensions of climate change in Bangladesh

The previous sections illustrate the state of inequalities in employment opportunities, including its geographical dimensions, and show how climate change contributes in making the gaps and deprivations. Women are the worst victims of these gaps and deprivations. A closer look at the facts may help to understand the situation more clearly. Male members are more engaged in service sectors (43.05 percent, followed by agriculture (41.81 percent)) that are mainly located in urban areas, where women are mostly engaged in agricultural activities (68.13 percent, followed by service sector (19.35 percent)) and the industry sector (12.52 percent). It suggests that females are more engaged in agricultural activities (26.32 percent point or 63 percent more than their male counterparts). This high level of engagement of women in agricultural activities indicates that they are forced to tolerate various forms of challenges that the agriculture sector face (see impacts of climate change on agriculture in Bangladesh (CCC 2009)) and have to take on additional workloads to keep the sector moving and at the same time protect their employment opportunities. On the other hand, male members, who are primarily engaged in the service sector, are less vulnerable since service sectors mainly occur in urban areas where life maintaining amenities and security/protection are more robust than rural areas. In rural areas facilities of health, education, water supply, communication, etc. are fragile and these conditions even cause the spending of more money from their small earnings. Moreover, in the wage rate women are less privileged than the men and it is worse in rural areas. Women get even less working hours of paid employment opportunities compared to their male counterparts. Table 9.5 shows that highest 45,000 women work 15–29 hours per week (among them 42,000, i.e., 93 percent come from rural areas), where 279,000 men work in this category per week (14 percent women against 86 percent men in the same category). Other categories also indicate similar inequalities in terms of working hours and availability, against which their earnings are accounted for.

Table 9.5 Employed youth (15-29) by major occupation, weekly hours worked, sex (in 000)

Occupation category		Hours worked group						
		Total	<15 hours	15-29	30-39	40-49	50-59	60+ hours
Bangladesh (covers 10 major occupational categories*)		19,342	255	4,416	1,007	6,468	4,550	2,645
Skilled agri and fisher workers	Total	1,354	83	324	224	343	269	111
	Male	1,248	63	279	219	318	258	110
	Female	106 (89 rural)	20 (16 rural)	45 (42 rural)	5 (5 rural)	24 (17 rural)	11 (10 rural)	01 (0 rural)

Note: Other categories are special occupation, legislators, senior officials and managers, professionals, technicians and associated professionals, clerks, service workers and shop and market sales workers, craft and related trade workers, plant and machine operators and assemblers, elementary occupations

Source: Compiled by Author

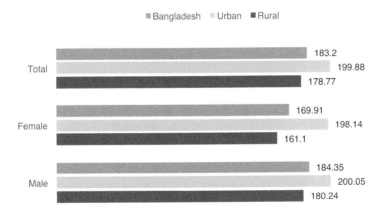

Figure 9.3 Daily wage rate in Taka

Source: Bangladesh Bureau of Statistics (BBS). 2010b. *Report of the Household Income and Expenditure Survey 2010*. Dhaka: Ministry of Planning

160 *Sheikh Tawhidul Islam*

Fewer working hours suggest two dimensions, firstly, employment opportunities are less in rural areas because of gaps in infrastructural and other supportive facilities like availability of energy; in certain cases, climate change and disaster impacts destabilize the employment sectors, which in turn lessen the employment opportunities as a whole in rural areas. Secondly, employers prefer men to women for certain reasons.

It is evident that during and post disaster situations men move elsewhere as they lose their places of works (e.g., agricultural land goes under water or becomes destroyed) and women shoulder the responsibilities to maintain households' sustenance (Nasreen 1995, 2012). Women have to shoulder the major responsibility of procuring and processing food and fuel, collection of drinking water, maintaining health and sanitation, looking after children and elderly, as well as livestock and poultry. However, the assessment indicates that (a) women are more engaged in climate sensitive sectors like agriculture than the men, (b) women working forces are mainly confined in rural areas where deprivations and insecurities are more prominent and, finally, (c) women endure more unequal conditions in terms of wage rate, availability of working hours and receiving basic service facilities.

5. Concluding remarks

The chapter reveals that the uncertainties and extreme events are natural parts of the climatic systems of the country where seasons are characteristically distinct. These uncertainties for the case of rainfall variable could be expressed in terms of heavy rainfall in some years, relative drought conditions in the other years. These irregular events occur in regular and rhythmic fashion in some cases and sometimes do not follow any pattern when putting these events in temporal scales. The second important aspect to mention about rainfall is that no significant historical change in the amount of rainfall occurrence was identified or no seasonal shift has been detected in this work. Mirza et al. (1998) also analyzed long-term annual precipitation records of meteorological sub-divisions of the Ganges, Brahmaputra and Meghna river basins and found no general significant change, with slight exceptions in a few meteorological sub-divisions. Shrestha et al. (2000) also carried out similar kind of work in Nepal. They did not find any distinct long-term trends in precipitation records at the 78 stations distributed across Nepal. McLean et al. (1998) and Lal and Aggarwal (2001) in this respect declared that there has been no discernible increasing or decreasing trend in annual precipitation in the greater Himalayan region during the last 100 years.

While considering temperature data it can be said that a general increasing trend in temperature for most of the weather stations of Bangladesh emerged. Most of the rise took place in winter minimum temperature and coastal stations show highest increase compared to the stations located in other parts of the country. The results indicate anomalies and variabilities exist in the climatic variables (i.e., temperature and rainfall) for almost all the stations of the country but do not provide a strong signal of change compared with the past. Not getting a strong pattern or signal of change may be attributed to the fact that many factors that influence climatic change of an area were not possible to include in the analysis. These are clouds/water vapor, wind systems, impacts of pressure gradient, location of the place within the broader region, land use and vegetation cover of the area, oceanic convection, biophysical, biological, bio-geochemical processes, effects of urban heat islands and various other human processes which need to be understood fully and incorporated in the assessments before making firm conclusions and predictions.

Despite having limited works undertaken in Bangladesh to create an in-depth understanding about the change in the climatic variables, government and non-government agencies, development partners, local government agencies and community based organizations have been trying to devise plans and strategies and implement programmes to reduce the climate change impacts and to enhance the resilience of the people and processes and systems of various kinds (e.g., economic, social, institutional, etc.). The Government of Bangladesh has been trying to address the climate change induced vulnerabilities mainly by implementing different social safety net programmes (Planning Commission, 2015). Vulnerable Group Development (VGD) is one such programme, originally designed and implemented by World Food Programme immediate after the 1974 famine conditions in Bangladesh, through which the Ministry of Women and Children Affairs (MoWCA) is supporting 750,000 ultra-poor women in all 64 districts of the country. Initially, the programme aimed at household level food security aspects (called Vulnerable Group Feeding (VGF)), which gradually extended its coverage and currently the programme addresses climate change related shocks in many ways. Even the government's annual expenditure to climate sensitive sectors currently stands at about 1 Billion USD (GED 2012), which plays an important role in reducing climate change related impacts and vulnerabilities.

References

Ahmed, R. and S. D. Shamsuddin. 2011. *Climate Change: Issues and Perspectives for Bangladesh*. Dhaka: Shahitya Prakash.

Bangladesh Bureau of Education Information and Statistics (BANBEIS). 2016. *Climate Change Education for Sustainable Development*. Government of the People's Republic of Bangladesh: Ministry of Education (MoE).

Bangladesh Bureau of Statistics (BBS). 2007. *Labour Force Survey 2005–06*. Dhaka: Ministry of Planning.

Bangladesh Bureau of Statistics (BBS). 2010b. *Report of the Household Income and Expenditure Survey 2010*. Dhaka: Ministry of Planning.

Bangladesh Bureau of Statistics (BBS). 2011. *Population and Housing Census 2011*. Dhaka: Ministry of Planning.

Bangladesh Bureau of Statistics (BBS). 2015. Bangladesh Disaster-Related Statistics 2015: *Climate Change and Natural Disaster Perspectives*. Dhaka: Ministry of Planning.

Bangladesh Bureau of Statistics (BBS). 2016. 'Bangladesh: Disaster Related Statistics 2015'. In *Climate Change and Natural Disaster Perspectives*. Dhaka: Ministry of Planning.

Brammer, H. 2014. *Climate Change Sea Level Rise and Development in Bangladesh*. Dhaka: The University Press Limited.

Centre for Environmental and Geographic Information Services (CEGIS). 2016. *Assessment of Sea Level Rise on Bangladesh Coast Through Trend Analysis*. Climate Change Cell, Department of Environment, Ministry of Environment and Forests, Government of the People's Republic of Bangladesh.

Climate Change Cell (CCC). 2009. *Characterizing Long-term Changes of Bangladesh Climate in Context of Agriculture and Irrigation*. Comprehensive Disaster Management Programme, Ministry of Food and Disaster Management, Bangladesh.

Dasgupta, S., F. A. Kamal, Z. A. Khan, S. Choudhury, A. Nishat. 2014. *River Salinity and Climate Change: Evidence From Coastal Bangladesh*. Study Conducted for World Bank, pp. 5–28.

Diffenbaugh, N. S., S. Martin and R. J. Trapp. 2013. 'Robust Increases in Severe Thunderstorm Environments in Response to Greenhouse Forcing'. *Proceedings of the National Academy of Sciences of the United States of America*, 110(41): 16361–16366.

EM-DAT. 2016. *The International Disaster Database*. Centre for Research on the Epidemiology of Disasters – CRED, Brussels.

General Economics Division (GED). 2012. *Climate Public Expenditure and Institutional Review (CPEIR)*. Bangladesh Planning Commission, Government of the People's Republic of Bangladesh.

Goodbred, Jr. S. and S. Kuehl. 2000. 'The Significance of Large Sediment Supply, Active Tectonism and Eustasy on Margin Sequence Development: Late Quaternary Stratigraphy and Evolution of the Ganges-Brahmaputra Delta'. *Sedimentary Geology*, 133(3): 228–230.

Climate change crisis in Bangladesh 163

Habib, I. 1995. *Essays in Indian History: Towards a Marxist Perception*. New Delhi: Tulika.

Islam, S. T. and A. Neelim. 2010. *Climate Change in Bangladesh: A Closer Look Into Temperature and Rainfall Data*. Dhaka: The University Press Limited.

Kuehl, S. A., M. A. Allison, S. L. H. Goodbred and Kudrass. 2005. 'The Ganges Brahmaputra Delta'. *Special* Publication –*SEPM*, 83: 414–418.

Lal, M. and D. Aggarwal. 2001. 'Climate Change and Its Impacts on India'. *Asia Pacific Journal on Environment and Development*,7(1).

McLean, R. F., S. K. Sinha, M. M. Q. Mirza and M. Lal. 1998. 'Tropical Asia'. In R. T. Watson, M. C. Zinyowera and R. H. Moss (eds.), *The Regional Impacts of Climate Change: An Assessment of Vulnerability*. Cambridge: Cambridge University Press.

Mikhailov, V. and M. Dotsenko. 2007. 'Processes of Delta Formation in the Mouth Area of the Ganges and Brahmaputra Rivers'. *Water Resources*, 34(4): 385–387.

Mirza, M. M. Q., R. A. Warrick, N. J. Ericksen and G. J. Kenny. 1998. 'Trends and Persistence in Precipitation in the Ganges, Brahmaputra and Meghna Basins in South Asia'. *Hydrological Sciences Journal*, 43(6).

Nasreen, M. 1995. *Coping With Floods: The Experiences of Rural Women in Bangladesh*. Unpublished PhD Dissertation, Massey University, New Zealand.

Nasreen, M. 2012. *Women and Girls: Vulnerable or Resilience?* Dhaka: University of Dhaka, Institute of Disaster Management and Vulnerability Studies.

Planning Commission, Ministry of Planning, Bangladesh. 2015. *7th Five Year Plan (Background Study)*. www.plancomm.gov.bd/7th-five-year-plan/(last accessed on 17 July 2015).

Rogers, K. G., S. L. Goodbred and D. R. Mondal. 2013. 'Monsoon Sedimentation on the Abandoned Tide-Influenced Ganges-Brahmaputra Delta Plain'. *Estuarine Coastal and Shelf Science*, 131.

Samenow, J. 2013. *Climate Change May Boost Violent Thunderstorms, Study Finds*. www.washingtonpost.com/news/capital-weather-gang/wp/2013/09/24/climate-change-may-boost-violent-thunderstorms-study-funds/(last accessed on 4 June 2016).

Shrestha, A. B., C. P. Wake, J. E. Dibb and P. A. Mayewski. 2000. 'Precipitation Fluctuations in the Nepal Himalaya and Its Vicinity and Relationship With Some Large Scale Climatological Parameters'. *International Journal of Climatology*, 20(3).

Soil Resources Development Institute (SRDI). 2010. *Saline Soils of Bangladesh*. Report prepared as part of SRMAF Project, p. 37.

Part III
Applied geography (combining human and physical geography methods)

10 Geographical dimensions of food security in rural Bangladesh

Mohammad Abu Taiyeb Chowdhury

1. Introduction

Food plays a crucial role in the agro-based economy of Bangladesh. About 58 percent of the income of the population is allocated to food (HIES 2007). Sustainable food production and achieving food security for all are among the major challenges that Bangladesh is confronted with. According to the Article 15 (a) of the Constitution of Bangladesh, it shall be a fundamental responsibility of the State to secure its citizens to the provision of basic necessities of life, including food. The Government of Bangladesh is firmly committed to achieve food security for all concerned. The National Food Policy (NFP) has been focusing on sustained increase in real cereal output – mostly rice – as one of the main components of national food security and availability (NFP 2006). Since independence in 1971, Bangladesh has made considerable progress in reducing poverty and food insecurity. Large-scale famine (the one that occurred in 1974) no longer poses a threat to the nation. The country has reached its self-sufficiency target in rice production in 2000 –a remarkable achievement given that the population has almost doubled since independence. Food aid has played an important role in the past, but with increased domestic production of food (Syem and Chowdhury 2011),imports have declined and the nature of food aid has shifted from general relief to focused development assistance, targeting the most vulnerable in the society. Despite these encouraging trends, there is still a long way to go before Bangladesh could be truly food secure (see the state of national food security – a context for sustainable development). The availability of and access to food produced domestically are now key issues affecting basic survival, nutrition, national security and stability, making sustainable agricultural growth[1] and rural development[2] vital to addressing these challenges (World Bank 2000, 2008). Food security has been and will remain a major policy issue for Bangladesh. This chapter focuses attention on the geographic distribution of food insecurity in rural Bangladesh. The thrust there has been to identify the regions of relative food insecurity (areas of socio-economic stress); the underlying objective being to provide a method for an ongoing comparative evaluation as a necessary planning tool for sustainable rural-regional development in Bangladesh. Following a brief discussion on the food security concept (under the title theoretical/ conceptual frameworks), the chapter presents an overview of the current state of food insecurity in Bangladesh from the context of sustainable development[3]. The chapter then describes the methodology employed in the area using Bangladesh as a case. A summary of mapping, including the profile of highly food insecure regions, is then presented, followed by options for using advanced multivariate statistical methods and the possibility of using Food Insecurity and Vulnerability Information Mapping Systems (FIVIMS) in this kind of situation. The chapter also includes suggestions for further readings and resources.

2. Conceptual frameworks

2.1 The concept of food security

The food security concept is directly proportional to population increase and is determined by their changing age structure and consumption behaviour as an effect of income (relative food prices, preferences and lifestyles). The right to food is a question of entitlement to food. According to the Universal Declaration Human Rights (UDHR), Article 25 (1) 'Everyone has the right to a standard of living adequate for the health and well-being of himself and his family', including food. The 'right to food' as a basic human right has formally been accepted during the World Food Summit (WFS) in Rome in 1996. Food security exists 'when all people, at all times, have physical social and economic access to sufficient, safe and nutritious food that meets their dietary needs and food preferences for an active and healthy life' (FAO 1998:2). The definition captures the complexity of the problem as food security encompasses many issues ranging from food production and distribution to food quality, preferences and health status of individuals. On the other hand, the food security concept is associated with food intake at the individual level and food availability at other levels, i.e., household, sub-national/ regional, national and global. So, there is no easy way of measuring food security in a single sequence. To simplify, a conceptual framework has been used, which distinguishes three distinct but inter-related components of food security or what (Maxwell 1996) are considered as elements – food availability, food accessibility and food utilization (Figure 10.1). Although food availability and accessibility are considered to be the joint determinants of food security, often they fail to recognize other elements of food security: utilization, stability, reliability, vulnerability and sustainability (Alamgir and Arora 1991).

Typically, food availability means that food needs to be available at the national or regional level. This is achieved through ensuring that domestic food production, food imports/aids and national food stocks are together sufficient to cover national requirements. Availability of food does not guarantee access to food by individual or household, but access to food is contingent upon there being food available at all geographic levels. Similarly, food security at one level does not guarantee food security at other levels. Even when aggregate food supplies are adequate, a number of factors may prevent poor individuals or households from accessing the food. Simply because they may not have access to land for own cultivation, their income levels may be too low to purchase the necessary food items from the market at the prevailing

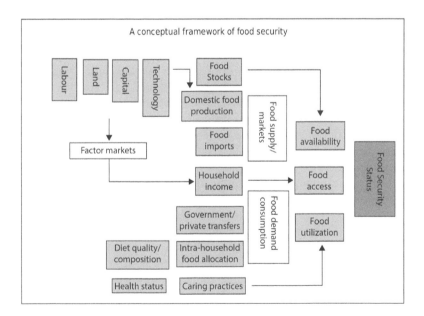

Figure 10.1 A conceptual framework of food security
Source: Adapted from the Food Security Atlas of Bangladesh 2004

price or may lack the necessary assets or access to credit to help them through difficult times. Furthermore, they may find themselves outside any public assistance or other programme that provides them with in-kind or cash transfers to supplement their food acquisition capacity (Alamgir and Arora 1991; WFP 2004). There is also a food distribution problem within a household: women, children, the elderly and the disabled often suffer from intra-household inequalities. It is also important how household members utilize the food. Often women eat last and least. Access to proper sanitation and health care, general nutritional awareness and caring practices are important determinants of an individual's capacity to absorb and utilize the nutrients in the diet and ultimately his/her food security status (Maxwell and Smith 1992; Chung et al. 1997).

2.2 Food insecurity

National availability of and household access to food alone are not sufficient to guarantee food security. Conversely, food insecurity is a human condition brought in by the inability to secure enough food to meet their minimum needs. This may be owing to the unavailability of food, lack of social or economic access, instability of supplies or income and/or inappropriate utilization at household level. Households most vulnerable to food insecurity generally include those which lack productive assets and depend on irregular income from daily wage labour – mostly casual jobs. Groups such as landless agricultural day labourers, casual fishermen and beggars fall into this category. Food insecurity leads to poor conditions of health and thereby poor nutritional status. Hunger, famine, malnutrition and undernourishment are different manifestations of the same reality. Since these are circular, closely intertwined and mutually reinforcing phenomena, they can be used as proxies for often food insecurity indicators (Chung et al. 1997)

There is an important geographic dimension to poverty, food insecurity and vulnerability in rural Bangladesh. These events have a disproportionate effect on people in marginal and risk prone areas, as poor people living in adverse peripheral locations are more likely to be vulnerable due to periodic climatic shocks. There are other households who are vulnerable because they are susceptible to natural disasters or are inaccessible or unsuitable for agricultural production. Similarly, homes of indigenous people often lack adequate services such as health care facilities, appropriate schooling and proper sanitation and are, therefore, frequently among the poorest of the country. Food insecurity also has a rural-urban dimension as most of the hardcore poor live in remote rural areas; a population dimension as regions with extreme poverty are more exposed to hunger and food insecurity; a gender dimension as children, adolescent girls, pregnant women and nursing mothers are especially vulnerable in a food insecure situation; a health dimension as the disabled and the elderly face relatively high nutritional risk (WFP 2004).

There is also a temporal dimension – defined as *transitory*, when a population suffers a temporary decline in food consumption, and as *chronic*, when a population is continuously unable to acquire sufficient food. Transitory food insecurity is associated with seasonal variation in food production and natural hazards, such as floods, cyclones and droughts. Depending on its asset endowments, a household adopts different strategies – e.g., borrowing money and the sale or consumption of productive assets –to cope with these events. These practices may undermine the long-term productive potential of poor households and may eventually lead to chronic food insecurity or complete deprivation (WFP 2004).

The root cause of food insecurity is poverty –a condition of those in an underdeveloped state of possessing fewer resources than required for providing oneself with the physical and conventional necessities of life. A comprehensive view of poverty recognizes that it is the denial of opportunity and choices most basic to human development (UNDP 1998),which challenges the common view that the shortage of food is the most important explanation of famine (Stanley 1998).

3. The state of national food security: context for sustainable development

Bangladesh is ranked 129th out of 169 countries in the 2010 Human Development Index (HDI) and ranked 70th among 122 countries in 2011 Global Hunger Index (GHI) (UNDP 2010; Chowdhury 2012; IFPRI 2012).

3.1 Technological issues and environmental challenges

The impressive growth of rice production over the past three decades or so has generated a sense of complacency regarding Bangladesh's ability to meet the growing demand for rice. However, the supply side of national food security upon further examination appears to be somewhat less promising (Hossain 1989; Ahmed 2001; Chowdhury 2009). Recent trends in the growth of rice production raise serious concern regarding the sustainability of the past achievements (Alam and Moral 2006). Intensive rice cultivation has been under strains since 1990. Soil degradation is undermining the long-term productive capacity of irrigated agro-ecosystems. Despite increases in input application, rice yields per acre have declined or stayed the same on about two-thirds of the areas planted with Modern High Yielding Varieties (MHYVs) in the 1990s (Pagiola 1995). Out of 27 agro-ecological zones in Bangladesh, 18 fall in the nutrient grades of poor to very poor (Ahmed and Hasanuzzaman 1998; Scherr 1999). Stress on land and water are further magnifying the issue of resource constraints to food production. There are also environmental issues: loss of paddy due to floods and natural disasters has become common phenomena, seriously disrupting the entire economy. Domestic food production remains susceptible to floods (Ninno et al. 2001) and droughts (Elahi 2001) of varying degrees, thereby perpetuating the threat of major production shortfalls, inadequate food availability and vulnerability from fluctuation in prices (Hossain and Shahabuddin 1997; Chowdhury 2009). In addition, there is well founded fear or threat of possible climate change. Bangladesh is one of the most vulnerable countries in the world to climate change impacts. The probable impacts of sea-level rise and the associated impacts on ecosystems and economic loss add to the already daunting array of environmental issues.

3.2 Development issues: social well-being and equity

There are equity issues as well in assessing national food security needs. Although Bangladesh has been making remarkable progress since 1971, these achievements are paradoxical so to speak that mask a grim reality –i.e., unevenness in the advances. Bangladesh produced nearly 500 kilograms of cereals per head in the year 2000, yet amid this abundance around one-third of the population was denied access to food and income inequality has been worsening (World Bank 2005). Poverty is the main obstacle in achieving food security (World Bank 2005). Around half the population lives below the established food-based upper poverty line (2,122 kcal/person/day) and a third remains below the lower poverty line (consumes fewer than 1,805 kcal/person/day – the minimum standard set by the World Food Programme (World Bank 2005; BBS 2012). With respect to food utilization, a large number of the Bangladeshi people do not consume sufficient food or survive on a diet lacking in micronutrients. Although cereal consumption among the population of Bangladesh has been increasing, undernutrition indicators remain alarmingly high and the rich/ poor gap is widening. The prevalence of malnutrition is also high in the country as a result of poor feeding habits and lack of access to nutritional food (World Bank 2005). There is also a problem of economic accessibility towards purchasing food; Bangladesh faces a difficult land distribution problem. According to 1996 Agricultural Census, about 29 percent of rural households did not own any land other than the homestead. About 49 percent owned less than 0.5 acres (functionally landless). The top 8 percent of households owned 48 percent of the total land in holdings of more than 5 acres (Rahman 1998; HIES 2007). There are more than 60 million food insecure rural households in Bangladesh. About 90 percent of the poor live in rural areas. In terms of land use changes, economic growth, industrialization, real estate development and rural-urban migration are leading to rapid urbanization. An important implication of rapid urbanization is that some of the fertile agricultural land has to be converted to other uses. The rice field has already started shrinking in Bangladesh (Halim and Rahman 2001).

4. Methodology: Bangladesh – a case application

This section presents a summary of an approach to, and results of, a comprehensive field investigation of food insecurity in Bangladesh, conducted at the Upazila (Sub-District) level. The research methodology has been developed in connection to the Food Security Atlas of Bangladesh, as the final product of a collaborative effort, prepared jointly by the Government of Bangladesh (Programming Division of the Planning Commission in co-operation with the Bangladesh Bureau of Statistics (BBS), Ministry of Planning) and the Vulnerability Analysis and Mapping Unit of the UN World Food Programme (UN-WFP).

4.1 Mapping food insecurity

Presenting geographic information in the form of a series of useful thematic maps provides an easy tool to assist in making spatial policy decisions. In the face of adversity and scarce resources, it is important and convenient to target the most deprived populations and vulnerable areas of the country to ensure the highest impact. It may be useful to stress here that where development occurs unequally across the country, regional differences in the relative levels of living are likely to become urgent political issues. Such is the case of Bangladesh that frustrates the development of society's potential (Chowdhury and Troughton 1989).

Indicators: Food insecure populations can be identified on a map by their geographic origin. Factors such as tendency to natural disasters, quality and distribution of agricultural land, access to education and health facilities, level of infrastructure development, employment opportunities and dietary and care giving practices provide possible explanations for this. The Food Security Atlas of Bangladesh, particularly the relative food insecurity map (2004), was based on the concept of food security as explained in the previous section (conceptual frameworks). It took into account the availability of food, access to food, utilization of food and the vulnerability factors. Food security issues were then identified for each of these four components through a comprehensive process. Subsequently, key indicators were selected and assigned to each issue (Table 10.1), providing a measure of the magnitude of the problem for every Upazila (Thana/Sub-District) in the country. The consultative process involved experts from government, UN, donor and NGO agencies.

Table 10.1 Key indicators of food insecurity

Main components	Major issues	Key indicators
Availability of food	• Low agricultural production	• Proportion of areas with bad cropping pattern • Proportion of households not owning agricultural land
Access to food	• Limited infrastructure/ market opportunities • High incidence of poverty • High seasonal unemployment	• Proportion of area within 2.5 km from major roads • Proportion of households having access to electricity • Proportion of households with pucca and semi-pucca houses • Proportion of households with non-agricultural source of income
Utilization of food	• Limited health care facilities • Limited education facilities • Unawareness regarding hygiene and sanitation practices	• Proportion of area within 50 minutes travel time to nearest hospitals • Proportion of households with heads having completed Grade 5 education • Proportion of households with pucca and semi-pucca sanitation
Vulnerability	• Frequent occurrence of natural hazards	• Disaster prone areas (flood, drought, erosion and cyclone)

Source: Adopted with modifications from the Food Security Atlas of Bangladesh 2004

Sources of data: Household level data were generated by the BBS based on a 5percent sample of the 2001 Population Census. Location specific indicators, such as cost-distance calculations, cropping patterns and disaster proneness were derived by WFP using the GIS-based datasets from the Local Government Engineering Department (LGED), and the Bangladesh Agricultural Research Council (BARC).

Analytic methods: In order to produce a ranking of the Upazilas in terms of their relative food security status, an index was derived based on the ten indicators presented in Table 10.1. Each indicator was standardized by ranking them from 1 to 511 (which equals to 470 Upazilas and 41 Urban Thanas. By summing up the rank values for each Upazila, a total score was derived. All 511 Upazilas were subsequently classified into four equal categories reflecting their relative levels of food insecurity: very high, high, moderate and low. The overall result is a marked clustering of areas classified as having a high or very high level of food insecurity. Six clusters are identified in the map as the food insecure regions (WFP and GoB 2004). In the following, a brief description of these geographic clusters is provided for general understanding.

5. Regional patterns in food insecurity: a planning perspective

5.1 Profile of highly food insecure regions

Food security in Bangladesh is characterized by remarkable regional variations. There is a gross disparity in regional food grain production and consumption (Syem and Chowdhury 2011). As the relative food insecurity map (2004) depicts (six clusters can be conveniently identified, see the Atlas, Figure 10.1), the most vulnerable food insecure regions in Bangladesh are the North-West Region, the Northern Chars, the Drought zone, the Sylhet Basin, the Coastal Belt and the Chittagong Hill Tracts (CHTs).

The North-West Region of Bangladesh is a slightly sloped plain at the foot of the Himalayan mountain range – traditionally, a food surplus production area. Despite this, a large percentage of the population remains food insecure, primarily because of the outcome of poor food access resulting from unequal land distribution, low agricultural wages and the impact of natural disasters, particularly susceptible to flash flooding during the monsoon.

The Northern Chars are unstable lands concentrated around the Brahmaputra and Jamuna rivers. These are very marginal lands, characterized by high level of flooding, riverbank erosion and instability. The sandy soil is not suitable for cultivation.

The Drought Zone is concentrated just north of the Padma River including the westernmost part of the Nawabganj, Rajshahi and Noagaon districts. A combination of high temperature, low annual rainfall and soil moisture deficiencies have a devastating impact on agricultural production, which threaten both small-scale farmers and labourers. Limited and expensive irrigation options exacerbate this problem further.

The Sylhet *haor* Basin is a low lying area that is under water for half of the year. Although some crops are grown during the winter season, most of the agricultural fields and much of the road network is submerged from May to October – the rainy season. Fishing is the main industry during the rainy season.

The Coastal Belt is part of an active delta including a number of small and large islands, as well as Char lands. Although a former granary of Bangladesh, the region has suffered from river erosion, saline intrusion, frequent natural disasters and out-migration. The remaining population is highly food insecure and isolated from reliable transportation and communication systems. This belt is highly prone to a number of natural hazards. Almost every year there is damage to crops, houses, livelihoods from floods or tidal waves. Cyclones are common in this area and occur on average every two to three years. The ultra-poor are generally most at risk, as their homes tend to be built with weaker materials and located on the most marginal lands.

The Chittagong Hill Tracts (CHTs) is a hilly, forested region in South-eastern Bangladesh – distinct from rest of the country. Most of the population (12 different tribal groups alongside a non-tribal population) is located in very remote rural areas making access to basic central services difficult. Much of the territory is composed of reserve forests; access to cultivable land is a key issue in the CHTs; most people in the region are agricultural labourers; the lack of access to cultivable land means very low food production, which has a great impact for household food security. The region is earthquake prone. Illegal and excessive logging has led to deforestation and soil erosion, increasing the region's vulnerability to landslides and reducing the availability of arable land. The smaller ethnic groups who live in the most remote and hilly areas, where it is most difficult to cultivate the land, often engage in traditional shifting cultivation (*Jhum*).

6. Strengths and limitations of the methodology used

The methodology used in the Food Security Atlas provides an analysis of the food security situation in Bangladesh using a series of thematic maps. These have been achieved through generating food security indices and translating those into user-friendly maps. Although the relative food insecurity map is easily understood and objectively verifiable, it needs to be used with caution. As the food insecurity map is based on a composite index of key indicators, it tells us neither the reason why a particular area is poor or vulnerable to food insecurity nor can it provide an answer regarding the best course of action to address the problem. In fact, it cannot tell us the magnitude of the problem with reference to any individual indicator/variable used in the analysis, for which a direct estimation of poverty (proportion of people below the lower poverty line) and malnutrition – stunting map and underweight map – were produced separately to help explain the spatial patterns. Underlying reasons for food insecurity are then explored following the concept of food security highlighting four components: (a) food availability: food supply, (b) food access: sources of income, infrastructure and markets, (c) food utilization: public health, water and sanitation, and (d) sources of vulnerability: natural disasters and household vulnerability. Geographic information on key factors is then presented with the expectation that these may contribute to regional variation in food security. It also includes an analysis of the variation in human capital – the population dynamics and migration patterns; description of the food security issues of the indigenous communities and those living in the inaccessible areas of Bangladesh; education and food security: primary and adult educational needs across regions. Tables are presented on key factors that may contribute to regional variation in food security. The food insecurity map points us towards a number of regions/areas that are likely to experience high food insecurity. The Atlas contributes significantly to a better understanding of food insecurity and vulnerability in the country (WFP 2004). In a nutshell, the analytic method used is satisfactory in itself. However, there is still ample scope to do more advanced analysis.

7. Concluding remarks

In recognition of the complex, multivariate nature of the problem of food security, two contrasting, yet complementary, advanced multivariate statistical analytical techniques may be suggested, i.e., trend seeking ordination (principal component analysis, factoring analysis, etc.) and recognition of group structure (i.e., hierarchic agglomerative clustering techniques) for possible future uses. These methods can be employed for the objective summarization of the data metrics as data reduction techniques. In this way, all the indicators/variables taken into consideration for the study could be analyzed in a single sequence. Instead of calculating a single composite index (as has been done in this case), a principal component/factor analysis could be utilized to identify major components/factors based on an underlying regularity in the dataset. Component/factor scores could then be used as input for further analysis – preparation of dimensional maps. A Dendrogram of the classification hierarchy could conveniently be constructed to recognize group structure (by sum of squares method of clustering – an agglomerative method, which hierarchically groups the cases, i.e., Upazilas so that those with greater similarity or lesser Euclidean distance cluster together in the numeric manipulation of the data). Details of the analysis could be found in the Statistical Package for Social Sciences (SPSS) or Statistical Analysis Systems (SAS). These techniques represent a sophisticated approach to identify the areas of socioeconomic stress or pockets of poverty (declining regions), multivariate socio-economic regionalization (creating development typologies) or as a measure of social well-being and equity. Among the suggested further readings and resources, the works of Berry 1960,1965; Smith 1972,1973; Abdullah 1979;Chowdhury 1984,1988; Chowdhury and Troughton (1989) are noteworthy.

There is also scope to update the Atlas. The Government of Bangladesh, together with the development partners, stakeholders, food, agriculture and nutrition scientists, human geographers, anthropologists, welfare economists and social policy planners, among others, can update the Atlas at regular interval (some prefer every three to five years). Hopefully, the next update version of the Atlas may include topics like urban versus rural food security, gender and food security, NGO coverage in the field of development interventions, government's targeted food aid programmes/WFP assisted programmes, and institutional capacity of the government to support the hardcore poor in crisis.

As an added dimension to the ongoing work, there is another possibility as well in food insecurity monitoring, analysis and mapping. One possibility is the use of FIVIMS. FIVIMS is a tool with which progress towards one of the most important Millennium Development Goals (MDGs) – the poverty and hunger goal – can be measured. FIVMS is a national system that aims to improve readily available food insecurity information in order to better utilize it for food security policies and improve action programmes. FIVIMS aims to build up a network of agencies that are committed to the FIVIMS objectives and support the Government of Bangladesh in improving food security policies based on reliable, timely and consistent data. It may be mentioned here that the Food Security Atlas is a contribution of the World Food Programme to the FIVIMS initiative. FIVIMS plays a significant role in the preparation of the Poverty Reduction Strategy Paper (PRSP) by providing the sub-national level data on extreme poverty, food insecurity and malnutrition. The Programming Division of the Planning Commission carries out the overall coordination of the FIVMS programme in Bangladesh. The FIVMS unit is located in the office building of BBS.

In order to reduce poverty, food insecurity and malnutrition, future efforts by the Government of Bangladesh and its development partners are more likely to be guided by the PRSP and the MDGs. The upcoming United Nations Summit for the adoption of the post-2015 development agenda will be held from 25 to 27 September 2015, in New York and convened as a high-level plenary meeting of the General Assembly. In that respect, the theme of this chapter is timely, and consistent with the first three Sustainable Development Goals (SDGs) of the United Nations: Goal 1 End poverty in all its forms everywhere; Goal 2 End hunger, achieve food security and improved nutrition and promote sustainable agriculture; and Goal 3 Ensure healthy lives and promote well-being for all at all ages. To reach the targets set in the SDGs, it will be necessary to target resources towards the most deprived and vulnerable regions/areas to ensure the highest impact. The Food Security Atlas of Bangladesh contributes significantly to a better understanding of food insecurity in the country and as such will prove to be a valuable tool in the programming of resources– including the Government and development partners, all concerned parties and individuals– working towards a poverty and hunger free Bangladesh.

To reduce poverty and food insecurity in Bangladesh, it is crucial to develop rural areas– where most of the poor people live. Of course, there is a question of rural-urban dichotomy in spatial decision-making. In this regard, Islam 2012 provides a more balanced view of development, justifying the necessity of a national land use plan. According to Islam,

> the future of Bangladesh needs to be visioned as a sound mix of urbanized areas and rural-agricultural spaces. This will require zoning of agricultural lands and very strictly maintaining its status as such. At the same time, forests, hills, and water bodies and wetlands must be conserved faithfully (CUS 2012: 15).

Developing rural areas in Bangladesh requires growth in rural economy- both agriculture and non-farm sectors grow faster. This can be addressed by adopting appropriate policy at the regional level, including institutional changes that are conducive to sustainable agricultural growth, while maintaining and sustaining the natural resources bases, e.g., land and water. Future agricultural development will need to focus more on organic farming, ensuring that soil retains vital nutrients, including protection of biodiversity. This should increasingly be based on crop rotation, diversification and agro-forestry, and greater benefits should be reaped from rainfed agriculture with water tolerant and drought resistant major cereal cultivars, including secondary food crops– coarse grains, pulses, roots and tubers. This, in turn, will require transforming rural institutions, including local government, and the expansion of rural infrastructure, empowering rural people to increase their income and welfare and improved coverage and quality of social services such as health and education.

Acknowledgements

As a courtesy, the author would like to extend his special thanks to a number of key people, among others, who have been behind the success of the Food Security Atlas of Bangladesh (2004). They are Mr. M. Fazlur Rahman, Secretary, Planning Division and Member of the Planning Commission of Bangladesh; Mr. AKM Musa, Director General, Mr. Abdur Rashid Sikder, Deputy Director General, BBS; Mr. Gopal Chandra Sen, Chief, Programming Division, Planning Commission; Mr. Siemon Hollema, WFP; Mr. Steve Haslett and Mr. Geoff Jones, Statistics Research and Consulting Centre of the Massey University, New Zealand. Thanks are also due to Tanwir Adib Chowdhury (author's eldest son) for constructing the flow chart – Figure 10.1.

Notes

1 Sustainable agriculture: A sustainable system is one that survives and functions over some specified time; it can be extended or prolonged over many generations (for a long time) rather than just a few years. As it pertains to agriculture, sustainable describes farming systems that are 'capable of maintaining their productivity and usefulness to society indefinitely. Such systems must be resource-conserving, socially supportive, commercially competitive, and environmentally sound' (Ikerd 1990). Sustainable farming is more a way of life to its practitioners than merely a law or regulation. At each step taken, it benefits farmers, brings welfare to their families, fosters community at large and helps preserve and protect the environment for future generations (United Nations 2009).
2 In general, rural development is conceived as a strategy to improve the socio-economic life of the rural poor. It involves extending the benefits of

development to the poorest in the rural areas, e.g., small farmers, tenants, landless and other disadvantaged groups. It is also concerned with the modernization and monetization of rural society and with its transition from traditional isolation to integration with the national economy. Rural development recognizes that improved food supplies and nutrition, together with basic services such as health and education, can not only directly improve the physical well-being and quality of life of the rural poor but can also indirectly enhance their productivity and ability to contribute to the national economy (Lipton 1977; ADB 1985; FAO 1985; World Bank 1997).

3 The term 'sustainable development' was first popularized by the World Commission on Environment and Development (WCED) in its 1987 report entitled 'Our Common Future'. The book is also known as the Brundtland Report, after the Chair of the Commission and former Prime Minister of Norway, Gro Harlem Brundtland. As defined in the Report, 'sustainable development is development that meets the needs of the present without compromising the ability of future generations to meet their own needs'. It contains within it two key concepts: (a) the concept of needs, in particular the essential needs of the world's poor, to which overriding priority should be given and (b) the idea of limitations imposed by the state of technology and social organization on the environment's ability to meet present and future needs (WECD 1987: 2, 43). The definition implies that a development path that is sustainable in a physical sense could theoretically be pursued even in a rigid social setting. Even the narrow notion of physical sustainability implies a concern for social equity, between generations, a concern that must logically be extended to equity within each generation (WECD 1987: 4).

References

Abdullah, A. H. 1979. *Socio-Economic Regionalization and Its Implications for Development Planning in Peninsular Malaysia*. Unpublished Ph.D. Thesis, Department of Geography, University of Michigan, Ann Arbor, Michigan, U.S.A.

Ahmed, Q. K. and S. M. Hasanuzzaman. 1998. 'Agricultural Growth and Environment'. In R. Faroque (ed.), *Bangladesh Agriculture in the 21st Century*, pp. 81–108. Dhaka: The University Press Limited.

Ahmed, R. 2001. *Retrospect and Prospects of the Rice Economy of Bangladesh*. Dhaka: The University Press Limited.

Alam, M. S. and M. J. B. Moral. 2006. 'Changes of Rice Productivity Under Modern Inputs and Sustainable Agricultural Production in Bangladesh'. *Journal of Bangladesh National Geographic Association (BNGA)*, 31(1 and 2): 103–132.

Alamgir, M. and P. Arora.1991. *Providing Food Security for All*. International Fund for Agricultural Development (IFAD), Studies in Rural Poverty No. 1. New York: New York University Press.

Asian Development Bank (ADB). 1985. *Rural Development in Asia and the Pacific*. Proceedings of the ADB Regional Seminar on Rural Development (1), Asian Development Bank (ADB), Manila, Philippines.

Bangladesh Bureau of Statistics (BBS). 2012. *Statistical Year Book of Bangladesh*. Dhaka: Ministry of Planning, Government of Bangladesh.

Berry, B. J. L. 1960. 'An Inductive Approach to the Regionalization of Economic Development'. In N. Ginsburg (ed.), *Essays on Geography and Economic Development*, pp. 70–107. Chicago: Department of Geography, University of Chicago.

Berry, B. J. L. 1965. 'Identification of Declining Regions: An Empirical Study of the Dimensions in Rural Poverty'. In W. D. Wood and R. S. Thoman (eds.), *Areas of Economic Stress in Canada*, pp. 22–49. Kingston: Industrial Relations Centre.

Centre for Urban Studies (CUS). 2012. *CUS Bulletin 60–62: On Urbanization and Development*, 60–62 (December 2011 and June 2012), p. 15. Dhaka: CUS.

Chowdhury, M. A. T. 1984. *Disadvantage and Development in Rural Canada: A Three-Phase Study of the Dimensions*. Unpublished M.A. Thesis/Dissertation, Department of Geography, Faculty of Graduate Studies, University of Western Ontario, London, Ontario, Canada.

Chowdhury, M. A. T. 1988. *Dimensions of Development and Change in Bangladesh, 1960–1980*. Unpublished Ph.D. Thesis/Dissertation, Department of Geography, Faculty of Graduate Studies, University of Western Ontario London, Ontario, Canada.

Chowdhury, M. A. T. 2009. 'Sustainability of Accelerated Rice Production in Bangladesh: Technological Issues and the Environment'. *Bangladesh Journal of Agricultural Research*, 4(3): 523–529, Gazipur: Bangladesh Agricultural Research Institute (BARI).

Chowdhury, M. A. T. 2012. 'Meeting Food Security Needs Through Sustainable Agriculture: A Perspective With Reference to Bangladesh'. *The Palawija*, 29(2): 1–6, UN-ESCAP Centre for Alleviation of Poverty Through Sustainable Agriculture (CAPSA).

Chowdhury, M. A. T. and M. J. Troughton. 1989. 'Dimensions of Change in Rural Bangladesh, 1960–1980'. *Rural Systems*, VII(1): 1–24, International Geographical Union (IGU) Commission on Changing Rural Systems' Official Journal.

Chung, K. et al. 1997. *Identifying the Food Insecure: The Application of Mixed-Method Approaches in India*. Washington, DC: IFPRI.

Del Ninno, C., P. A. Dorosh, L. C. Smith and D. K. Roy.2001. *The 1998 Floods in Bangladesh: Disaster Impacts*. Household Coping Strategies and Response, Research Report No. 22. Washington, DC: IFPRI.

Elahi, K. M. 2001. 'Drought in Bangladesh'. In K. Nizamuddin (ed.), *Disaster in Bangladesh*. Dhaka: Department of Geography and Environment, University of Dhaka.

Food and Agricultural Organization (FAO). 1985. *Toward Improved Multilevel Planning for Agricultural and Rural Development in Asia and the Pacific*. FAO Economic and Social Development Paper 52. Rome: Food and Agricultural Organization of the United Nations (FAO).

Food and Agricultural Organization (FAO). 1998. *The Right to Food: Theory and Practice*. Food and Agricultural Organization of the United Nations. Rome: FAO Information Division.

Halim, A. and Z. Rahman. 2001. 'Agriculture'. In *Bangladesh Environment Outlook 2001*, pp. 131–147. Dhaka: Centre for Sustainable Development.
HIES. 2007. *Household Income and Expenditure Survey 2007*. Dhaka: Ministry of Planning, Bangladesh Bureau of Statistics (BBS).
Hossain, M. 1989. 'Food Security, Agriculture and the Economy: The Next 25 Years'. In *Food Strategies in Bangladesh: Medium and Long Term Perspectives*. Bangladesh Planning Commission. Dhaka: The University Press Limited.
Hossain, M. and Q. Shahabuddin. 1997. *Sustainable Agricultural Development in Bangladesh: Challenges and Issues*. Paper presented at the Bangladesh-IRRI Dialogue, Dhaka, Bangladesh, 24–27 February 1997.
Ikerd, J. 1990. 'Sustainability's Promise'. *Journal of Soil and Water Conservation*, 45(1): 4.
International Food Policy Research Institute (IFPRI). 2012. *Global Food Policy 2011*. Washington, DC: International Food Policy Research Institute (IFPRI).
Islam, N. 2012. 'Towards an Urbanized Bangladesh: Looking Beyond 2050'. In *CUS Bulletin on Urbanization and Development*, pp. 60–62, June and December 2011, June 2012, pp. 13–15. Dhaka: Centre for Urban Studies.
Lipton, M. 1977. *Why the Poor Stay Poor: Urban Bias in World Development*. London and Canberra: Temple Smith Australian National University Press.
Maxwell, D. G. 1996. 'Measuring Food Insecurity: The Frequency and Severity of Coping Strategies'. *Food Policy*, 21(3): 291–303.
Maxwell, S. and M. Smith. 1992. 'Household Food Security: A Conceptual Review'. In S. Maxwell and T. Frankenberger (eds.), *Household Food Security: Concepts, Indicators and Measurements*. Rome and New York: IFAD and UNICEF.
National Food Policy (NFP). 2006. *National Food Policy Plan of Action (2008–2015)*. Dhaka: Ministry of Food and Disaster Management, Government of Bangladesh.
Pagiola, S. 1995. *Environmental and Natural Resource Degradation in Intensive Agriculture in Bangladesh*. Environment Department Paper No. 15. Washington, DC: World Bank.
Rahman, H. Z. 1998. 'Rethinking Land Reform'. In Faruque (ed.), *Bangladesh Agriculture in the 21st Century*. Dhaka: The University Press Limited.
Scherr, S. J. 1999. *Soil Degradation: A Threat to Developing-Country Food Security by 2020?* (Vol. 27). Washington, DC: International Food Policy Research Institute.
Smith, D. M. 1972. 'Towards a Geography of Social Well-being: An Inter-State Variation in the United States'. In R. Peet (ed.), *Geographical Perspectives on American Poverty*. Antopode Monograph Series in Social Geography, No. 1, pp. 17–46. Worter: Clark University.
Smith, D. M. 1973. *The Geography of Social Well-Being in the United States: An Introduction to Territorial Social Indicators*. New York: McGraw-Hill.
Stanley, J. 1998. 'Understanding Development: The Experience of Voluntary Agencies in South Asia'. *Asian Affairs*, 29(2): 152–159.

Syem, M. A. and M. A. T. Chowdhury. 2011. 'Regional Patterns of Food Grain Production in Bangladesh: Implications for National Food Security and the Environment'. *Asian Profile*, 39(4), August: 319–334.

United Nations. 2009. *Sustainable Agriculture and Food Security in Asia and the Pacific*. Bangkok: UN-Economic and Social Commission for Asia and the Pacific (ESCAP).

United Nations Development Programme (UNDP). 1998. *Overcoming Human Poverty*. New York: United Nations Publications.

United Nations Development Programme (UNDP). 2010. *Human Development Report 2011*. New York: Oxford University Press.

WFP and GoB. 2004. *The Food Security Atlas of Bangladesh*. Dhaka: World Food Programme and Ministry of Planning, Government of Bangladesh.

World Bank. 1997. *Rural Development: From Vision to Action*. Washington, DC: The World Bank.

World Bank. 2000. *Bangladesh: A Proposed Rural Development Strategy*. Dhaka: The University Press Limited.

World Bank. 2005. *World Development Report*. New York: Oxford University Press.

World Bank. 2008. *World Development Report: Agriculture for Development*. Washington, DC: The World Bank.

World Commission on Environment and Development (WECD). 1987. *Our Common Future*. New York: Oxford University Press.

11 Development induced displacement and resettlement practice in Bangladesh

Hafiza Khatun

1. Introduction

Compulsory displacements, which occurs for development reasons, embody a perverse intrinsic contradiction in the context of development (Cernea 2000). They raise ethical question because they reflect on the inequitable distribution of development's benefits and losses. For the last three decades, about ten million people worldwide have been displaced annually as a result of development projects and over four-fifths of this population is from East and South Asia (Wet 2004). On the other hand, since independence, large infrastructure projects have been implemented in Bangladesh, which has led to the displacement of about 0.05 million people annually (Khatun 2000b). More than 1.5 million people have been involuntarily displaced during the period of 1994 to 2016 for acquisition of more than 10,000 hectares of land for development partners' funded development projects, such as the Jamuna Multipurpose Bridge Project (JMBP), the Regional Road Management Project (RRMP), the Bhairab Bridge Project (BBP), the Jamuna Bridge Access Road Project (JBARP), the South-West Road Network Development Project (SRNDP), the Jamuna-Meghna River Erosion Mitigation Project (JMREMP), the Coastal Embankment Improvement Project (CEIP-I), the Bangladesh India Electrical Grid Interconnection Project (BIEGIP), the Padma Multipurpose Bridge Project (PMBP), the Tongi-Bhairab Double Line Railway Project (TBDLRP), the Chittagong Hill Tracts Rural Development Project (CHTRDP), and so on. While most of these projects are funded by Development Partners (DP) such as the World Bank (WB), the Asian Development Bank (ADB), the Department for International Development (DFID) and the Japan International Coopery (JICA), the Government of Bangladesh (GoB) has provided grants for others. However, some projects are being designed and implemented by following Public-Private Partnership

(PPP) investment initiatives of the Government of Bangladesh and other national/international companies; Dhaka Elevated Expressway Project (DEEP) is one of them. Project-affected people rarely, if ever, regain their lost ways of life and livelihood and often become impoverished as a result of being resettled in spite of international development partners having sets of resettlement guidelines and processes.

In Bangladesh, the GoB acquires land for any development project for 'public interest' by using 'eminent domain' under the Land Acquisition Ordinance of Bangladesh known as the Acquisition and Requisition of Immovable Property Ordinance 1982 II (ARIPO). This machinery (ordinance) of the GoB neither has any provision for Resettlement and Rehabilitation (R&R) nor any mechanism to assist and rehabilitate those affected and enable them to overcome displacement-led impoverishment. It merely provides for monetary compensation, that is, the Cash Compensation under Law (CCL) for legally owned property only. According to Cernea (2000),the major impoverishment risks for induced displacement without proper mitigation measures are landlessness, joblessness, homelessness, marginalization, increasing morbidity, food insecurity, loss of access to common property resources and social disarticulation. All these risks lead to uncertainty in the restoration of livelihood, at least to the pre-project level. DPs have developed various guidelines for resettlement planning by incorporating Cernea's (1996, 1998) model of Impoverishment Risk and Reconstruction (IRR) to turn risk into development opportunities. The WB policy statement on resettlement issues in 1980 was the first explicit recognition of this need and included in the project operational directives. Over the period of time since then, other DPs adopted their own policy and WB has also revised its own several times through experience of implementation in different types of project in different geographical locations. Experts believe that adverse outcomes of resettlement can be avoided or minimized by risk reversal through appropriate resettlement planning and adoption of proper mitigation measures detailed in Cernea's IRR model (Khatun 2009). This can be achieved through:

(a) Good policy and management practice
(b) Consultation with stakeholders
(c) Socially responsible design to minimize impact
(d) Good planning
(e) Appropriate livelihood restoration planning
(f) Enhanced capacity and commitment of the government/executive agency

The DPs' guidelines have suggested for appropriate resettlement and rehabilitation planning to avoid or minimize the created impoverishment of all the development induced displaces. In the mid-1990s, WB's guidelines were followed for the first time in Bangladesh, in the JMBP, very successfully. The project Affected Persons (APs) enjoyed R&R with unique examples through various kinds of mitigation measures, by following DPs' guidelines. Since then the GoB has been preparing and implementing a Resettlement Plan (RP) as an integral part of the DP funded projects by focusing on resettlement and poverty issues and gradually developing various techniques or methodologies appropriate for implementation. Special attention has been given to protect the livelihood and impoverishment of the disadvantaged groups (irrespective of legal entitlement on the affected property). However, 50 percent or slightly more of the APs of different projects could be able to overcome the impoverishment hazards even with these DP guided mitigation measures. The impoverished APs were mostly poor and marginal landholders prior to expropriation of their property. This is more specifically true for the agriculture base communities (Khatun 2000a; Khatun 2005). On the other hand, the development project acted positively for the community dependent on secondary or tertiary economic activities. This seems to be an inevitable consequence and could only be addressed under the resettlement and rehabilitation programme considering the spatio-temporal and social aspects. However, though slow paced, the percentage of APs with success stories has been on a rise over the last two decades (2000–2019). About 15 percent of the APs have not only overcome the impoverishment but also improved their livelihood. Over time, the percentage of APs with success stories is increasing gradually, although very slowly (ADB 2003a; Khatun 2015).

This chapter evaluates existing GoB LA Ordinance, DPs' requirement, development of strategies as 'good practices' for land acquisition, resettlement and rehabilitation for infrastructure projects in Bangladesh, with special attention to Jamuna Multipurpose Bridge Project.

2. Legal framework of land acquisition for development projects in Bangladesh

The first law on the land acquisition in the sub-continent was promulgated in 1870. This law was amended by the Land Acquisition Act, 1894 (Act I of 1894). While the land acquisition Act of 1894 remained enforced, the East Bengal Requisition of Property Act was promulgated in 1948 after the partition of India. The 1948 Requisition of Property Act was extended from time to time on a triennial basis and finally modified by the Ordinance of 1982 and ARIPO was enacted in 1982. This ordinance had subsequent amendments in 1993 (increased premium value on calculated market value from 20 to 50 percent) and 1994 (Payment for crop compensation to the *bargader* (share cropper/tenant farmers).

Once the government decides to acquire a piece of land for 'public purpose', the ARIPO is applied. Deputy Commissioner (DC) will act as the Acquiring Body (AB) and be vested with the power to acquire, determine and pay compensation for acquiring assets. The CCL is being paid by the DC office by assessing the market value (MV) of the lost assets (assessed on the basis of an average recorded value of similar kind of land in the vicinity for the preceding 12 months), plus 50 percent premium on assessed MV for compulsory acquisition of the land. Usually, the compensation amount is much lower than the Replacement Value (RV), that is, the actual market price of the acquired asset at the time of acquisition. In most of the cases, this amount of money covers less than 25 percent of the market value of the lost property (Khatun 2009). The RV is assessed through market surveys and information collected from various sources at the time of designing the RP. The land acquisition is always viewed with concern and apprehension by the APs, because it potentially diminishes the productive base of farm families, and because it is always associated with low and delayed compensation payment and often harassment from DC office (Zaman 1996). The ARIPO does not consider the socio-economic status of the affected persons prior to and after the commencement of the project.

3. DP requirements for resettlement

The R&R policies of all DPs are similar in nature. They advocate providing compensation for all the losses and also giving support to restore livelihoods and thereby reduce the risk of impoverishment among those displaced by development projects. Special attention is also given to protect the livelihoods of the disadvantaged groups (irrespective of their legal entitlement to the affected property). This is being implemented by the government executive agency (EA) through experienced non-governmental organizations (NGOs) to ensure that the project is beneficial for all the stakeholders. However, so far, in comparison to the entitlements under the GoB funded projects, the compensation package for the APs under DP funded projects is very generous. Table 11.1 illustrates the gaps between the eligibility for compensation under the ARIPO and the R&R policy of the DPs.

The compensation package offered under GoB and DP funded projects discriminates APs of the two types of project. At times, it causes

Table 11.1 Types of losses eligible for compensation under the Development Partners' Policy and the ARIPO of the GoB

Main components	Major issues	Key indicators
Availability of food	• Low agricultural production	• Proportion of areas with bad cropping pattern • Proportion of households not owning agricultural land
Access to food	• Limited infrastructure/market opportunities • High incidence of poverty • High seasonal unemployment	• Proportion of area within 2.5 km from major roads • Proportion of households having access to electricity • Proportion of households with pucca and semi-pucca houses • Proportion of households with non-agricultural source of income
Utilization of food	• Limited health care facilities • Limited education facilities • Unawareness regarding hygiene and sanitation practices	• Proportion of area within 50 minutes travel time to nearest hospitals • Proportion of households with heads having completed Grade 5 education • Proportion of households with pucca and semi-pucca sanitation
Vulnerability	• Frequent occurrence of natural hazards	• Disaster prone areas (flood, drought, erosion and cyclone)

Sources: Adapted with modification from the Food Security Atlas of Bangladesh 2004; compiled by author

serious difficulties for the implementing agency at field level, where they are responsible for the management of both DP and GoB funded projects in the same location. For obvious reasons, land acquisition by the GoB funded projects is considered as a curse or misfortune and almost a complete loss for the APs. So, uniform time and location demanding appropriate mitigation measures need to be devised for resettlement, rehabilitation and restoration of livelihood of all the victims regardless of source of funding of the project. In response to this demand, the GoB has prepared a draft national policy on involuntary resettlement in 2007, but it is yet to be enacted.

It is to be noted that irrespective of the compensation package, resettlement of displaced people is always disadvantageous. They rarely regain their lost ways of life and livelihood in spite of the mitigation measures (Zaman 1996; Cernea 2008). The percentage of APs with a record of regaining or changing their livelihood in the positive sense is not very encouraging in Bangladesh (ADB 2003b; Khatun 2009). The issues of resettlement, rehabilitation and livelihood restoration needs urgent attention at the design and planning stage so that appropriate and adequate mitigation measures can be factored in alleviating the induced impoverishment on the APs.

4. Current resettlement practice in Bangladesh

The DPs' concept of resettlement and rehabilitation of APs had added a new dimension to the practice of land acquisition and resettlement in Bangladesh. This has been done by preparing Land Acquisition and Resettlement Plan (LARP)/Resettlement Action Plan (RAP)/RP by incorporating additional entitlements on top of the legal entitlements recognized by ARIPO (GoB) only for respective DP funded project. The RP delineates the entitlement matrix and also outlines the mechanism of implementation, monitoring and the involvement of agencies, both governmental and non-governmental, and stakeholders to accommodate GoB and DP policies. So far a number of RPs have been prepared, approved, financed and implemented by the GoB for the infrastructure development projects. However, the GoB is financing the entire R&R component for all the projects irrespective of the source of funding agency for the project. For example, in the case of the JMBP, out of the total project cost of 960 million USD, the GoB contributed 350 million USD, which included the resettlement cost of 14 million USD. The R&R cost was about 7 percent of the total project cost. Over the period of time, [*since 1995 till today for every project*] the R&R cost in terms of percentage of total project cost is in a hiking trend. However, in recent years, due to DPs' policy requirements and its persistence to mitigate project induced displacement, some improvements in the form of 'good practices' are noticeable in most of the DP funded projects.

5. Good practices in dealing with involuntary resettlement issues

The JMBP and subsequent DP funded projects have established some 'good practices' in dealing with involuntary resettlement issues over the last two decades, which are as follows:

- Steps were taken to minimize involuntary resettlement by exploring all viable alternative design option. For example, out of four, the present location of Padma bridge was chosen as the best after considering all the options including resettlement. The JMBP experience guided the other subsequent projects to acquire optimum/minimum possible land
- Under projects where displacement was unavoidable, resettlement measures were conceived and implemented as a 'development programme'
- Accurate Census, Socio-Economic Survey (SES) aided with video-filming of structures and all assets before the cut-off date, land market and property assessment surveys were carried out to assess the replacement value of the lost assets. Video-filming facilitates in preventing fraudulent claims and was initiated in the BBP in 1997, later on it is being practiced in all the projects including the GoB funded projects
- All APs including those without title to land were identified and the entitlements of their losses were assessed at replacement rates
- Information about the planning and implementation of RPs were disseminated to all the stakeholders through appropriate mechanisms
- All primary and secondary stakeholders, including vulnerable groups, were involved in the consultation process through household surveys, focus group discussion, public consultations and formal meetings
- A time-bound RP, which identifies categories of APs and their entitlements, was prepared. The responsibilities of the implementing agencies, their budget and monitoring mechanisms were outlined clearly
- 'Identification Cards' (ID cards) with photo were given to APs with details of all their entitlements. Issuance of such cards gave confidence to those affected and enabled a proper tracking and monitoring of the R&R process
- Where relocation was required, relocation plans and options in consultation with APs and host communities were developed by synchronizing the implementation stages of the project. For example, two and five resettlement sites (RSs) have been established in the JMBP and PMBP, respectively, mainly for housing. A resettlement market was established in the BBP. The APs were given option of selecting either the developed sites or the host village
- 'Incentive driven mechanisms' were introduced for the host communities for increasing the carrying capacity of the village by providing/improving infrastructure such as roads, school buildings, mosques,

temple, tube-wells, and so on, so that they do not show resentment in accepting, rather welcoming the APs
- Special attention was given to the vulnerable APs, especially to the female headed, disabled headed households and the households under the poverty level
- Special attention was given to socially excluded community by proving additional facilities
- Appropriate income restoration programmes were established with the objective of restoring/improving the productive base and affected/existing living standard of the APs. Credit facilities were provided to all the APs of the JMBP through NGOs. A grant of Bangladesh Taka 10,000 in two instalments was granted to all vulnerable APs of the BBP
- A social development programme is being designed in the project area to cover primary as well as secondary APs; the PMBP is one of the example
- Different committees were formulated to facilitate the APs to participate in the implementation process of RP. This way the confidence of getting justice enhanced among the APs
- A non-government Organization (NGO), which has experience working with the grass root level people, was hired by the executing agency (EA) to implementing the RP
- The Common Property Resources (CPRs) of the community were restored /reconstructed
- Periodic monitoring and evaluation of resettlement operation were undertaken

Presently, the GoB does not have any resettlement policy in hand but nowadays different infrastructure implementing agencies are adopting various measures to accommodate ARIPO as well as DPs' policies with the goal of facilitating the APs to rehabilitate them to at least the pre-project condition. Based on the identification of APs and the assessment of RV of the affected assets through market survey and an agreed resettlement framework between GoB agency and DP, the assessment of compensation and valuation of the losses are done. When RV is higher than the CCL, additional money is paid by the project and known as a top-up. Table 11.2 illustrates a summary of a typical entitlement matrix with category of losses, entitlements and compensation package and potential beneficiaries as presented in a RP of a DP funded project. Some other common entitlements are a Transfer Grant (TG), a Reconstruction Grant (RG) and some other grants. Here all the affected persons are identified irrespective of legal

Table 11.2 A typical compensation package of a Development Partner funded project summary

Category of losses	Persons entitled/potential beneficiaries	Entitlements/compensation package
Agricultural/homestead, commercial, industrial land; Loss of water bodies (ponds, both cultivated and non-cultivated) and common property resources	Legal owner/title holders	CCL by DC Top-up Stamp duty and registration cost after purchasing new land. Total value up to RV
Residential, commercial/industrial structures with title to land	Legal owner/title holders	CCL Top-up TG and RG Taking away salvageable materials free of cost
Residential and other physical structures without title to land (squatters)	Socially recognized owners of structures identified during census	Compensation as per PWD, plus 50 percent Top-up TG and RG Taking away salvageable materials free of cost
Common property resources (CPRs) structures (a) with or (b) without title to land	Legal owners (or registered committees) Socially recognized owners of structures identified in census	CCL TG RG (a) Taking away salvageable materials free of cost. OR (b) Replacement value TG Taking away salvageable materials free of cost

(*Continued*)

Table 11.2 (Continued)

Category of losses	Persons entitled/potential beneficiaries	Entitlements/compensation package
(a) Trees with title to land and (b) Trees on public land or lessees	Legal owner Socially recognized owners of trees grown on public or other land, identified by census	(a) Timber trees and bamboos: CCL Top-up For fruit trees: CCL Top-up Market value of fruits for average of three annual years production OR (b) Timber trees and bamboos: Compensation for lost trees For fruit trees: compensation for lost trees Market value of fruits for average of three annual years production Allowed to take the trees free of cost
Standing crops/fish stock with (a) title to land (b) without title to land	(a) Legal owner cultivators (b) Socially recognized users of land	(a) CCL Top-up OR (b) Compensations for crops/fish stock. Allowed to take crops and fish stock
Leased or mortgaged agricultural land or ponds and commercial land	Titled and non-titled leaseholders/ licensees/share croppers	Cash grant to the share croppers, licensees and lessees of agricultural land, pond Allowed to take the crops/fish
Income from dismantled commercial and industrial premises	Proprietor or businessman or artisan operating on premises, at the time of issuance of notice u/s 3 by DC and/ or identified by census	Cash grant for loss of business income
Income (wage earners in agricultural, small business and industry)	Regular employees/wage earners identified by census	Cash grant equivalent to three months average income

Income from rented-out and access to rented-in residential and commercial premises	Legal owner of the rented-out premises	Grant for loss of rental income
Reconnection of utilities (gas, electricity, telephone, water, sewage, etc.)	Legal subscriber as Identified by DC	Cash grant for new utility connections (a) Gas (b) Electricity (c) Telephone (d) Water (e) Sewage
Assistance to vulnerable households	Households under the poverty level, head of household are elderly, disabled and very poor	One time grant in addition to other compensations
Assistance to poor female-headed households	Female-headed household under the poverty level	One time grant in addition to other compensations
Livelihood improvement Programme	One member of each vulnerable household and households losing 10 percent over of their total income	IGA Training 'seed grant' to each trained member for investment
Personal finance management programme	All households losing structure/trees/crops/land	Credit/grant
Unforeseen adverse impacts	Households/persons affected by any unforeseen impact identified during RP implementation	Compensations/allowance and assistance depending on type of loss
Temporary impact during construction	Households/persons and/or community affected by construction impacts	The contractor shall bear the cost of any impact All temporary use of lands outside proposed are returned to owner, rehabilitated to original, preferably better standard

Source: Compiled by Author

CCL – Cash Compensation under Law, DC – Deputy Commissioner, RV – Replacement Value, TG – Transfer Grant, RG – Reconstruction Grant, PWD – People with Disabilities, IGA – Income Generating Activities ownership of the acquired land. Social dislocation of community or indirectly affected people and dislocation of social/community infrastructures are also included in this identification list.

6. Salient features of the JMBP

6.1 Types of compensations paid and resettlement

Compensation was given for expropriation of all kinds of assets including land, structure, tree, crops, fish and so on owned by the occupier at present market rate irrespective of legal ownership of land but dislocated by the project in Bhuapur and Shirajgong districts.

To facilitate resettlement, rehabilitation and livelihood restoration, additional mitigation measures were included, covering payments for loss of income, monetary compensation for R&R that included expenses for relocation, reconstruction, purchase of alternate land and provision of other basic amenities and so on.

Apart from these, the JMBP provided short-term/long-term rehabilitation measures that comprised of training, Human Resource Development (HRD), Occupation Skill Development (OSD) and credit facilities through assigned NGOs that gave credit on terms and conditions similar to other credit-providing NGOs in the neighbouring areas.

6.2 Resettlement process

The implementation of the JMBP started in 1993 and was completed in 2000. For immediate relocation of the losers of homestead, the JMBP followed the twin strategy that:

(a) Encouraged the APs to relocate by themselves in the host villages surrounding the project area. The traditional practice of relocating housing structures affected by riverbank erosion in the vicinity was also followed
(b) Provided with an alternative option of relocation to RSs where community facilities including schools, hospitals, mosques, roads, electricity and water supply were developed for the vulnerable APs who could not relocate by themselves

After being compensated for the appropriation of homestead by the DC, all salvageable materials were returned to the APs free of cost. Each relocated Household (HH) was given a TG, an RG, a top-up

Development induced displacement 197

and stamp duty (SD) to buy an alternate land. Each household was provided with sanitary facilities as well. Squatters were given 101 sq. metres of homestead land in RSs free of cost. They could also buy another 101 sq. metres of land adjacent to their plots. At the RSs, schools, health centres, mosques and community centres, access to drinking water were provided at project cost. Besides, at least one member of the family was entitled to training, employment and income generation programmes. The displaced families were also given credit support through locally active NGOs.

To promote social and cultural integration of the APs with their host villages, a number of community infrastructures like roads, schools, mosques, temples, tube-well and so on, were developed in the host villages. This effort reduced potential conflicts between the APs and their hosts and developed a sense of belonging for the APs amidst the host community. Above all these facilities helped the host village to increase their carrying capacity.

6.3 Resettlement sites

The JMBP developed two RSs, initially to accommodate 755 households on either side of the river. Over time 1,200 plots were developed in a total area of about 140 hectares. Initially, people were not interested in relocation at the resettlement sites as it was a new concept. Besides, they also felt that the sites did not have the necessary infrastructural facilities, but, over time with the development of infrastructure and community facilities, they got well-adjusted to their new sites. The plot size varied from 101, 202 and 404 sq. metres per household (ADB 1999).

The plots at the resettlement sites were sold at par with the compensation price paid (CCL) at the time of land acquisition, on a 99-year lease arrangement. The sale of land at the resettlement sites was banned for 10 years from the date of purchase. For squatters and uthulies (constructed house structure on others' land with permission), the plot was jointly owned by the husband and wife and registered in their named and on separation (if any) the wife could retain ownership.

6.4 Other mitigation measures

A programme for Erosion and Flood Affected Persons (EFAPs) was undertaken to help people living in an area 8 km upstream and 10 km downstream of the bridge and compensation was provided for the loss of land, crops and employment due to erosion caused by flooding. This programme was initiated to compensate more than 180,000 people and was implemented by an NGO named Bangladesh Rural Advancement Committee (BRAC). The women headed households under the EFAP Programme were given a 20 percent additional relocation grant.

Another programme undertaken was the Environment Management Action Plan (EMAP). The aim of this programme was to mitigate the adverse impact on livelihoods of the people, fisheries and agriculture and wildlife of the area. In total, eight NGOs including the RP implementing NGO and a GoB organization, the RU of the JMBA, were involved in the project area. The NGOs were involved mainly for providing credit, raising awareness about human development and gender equity and enabling the affected people to adopt livelihood activities such as fish cultivation, poultry, cattle rearing and small business activities.

7. Concluding remarks

GoB land acquisition ordinance does not deal with broader social and economic impacts of land acquisition including compensation at replacement value, relocation or other assistances to overcome induced impoverishment. However, since the 1990s, the DPs' guidelines on R&R by funding agencies are being followed in the design and implementation of DP funded infrastructure projects in Bangladesh. However, in most of the projects, various mitigation measures are undertaken to compensate those affected, but logistical support such as credit, training and counselling are not provided.

The JMBP can be considered as an example of good resettlement in a rural setting and can be replicated as a case of best practice. The resettlement model of the JMBP can be improved by adding good practices from other projects that focus on the creation of employment opportunities and consult the APs in the designing of skill and training programmes. Resettlement plans should incorporate locally oriented measures, mainly for income restoration, undertaken so far in donor-funded projects are more effective than entirely government-aided projects.

The issue of an unequal compensation package offered to APs of GoB/private and DP-funded projects have been addressed in the draft National Policy on Resettlement and Rehabilitation (NPRR) of August 2007 prepared by the GoB with the Technical Assistance (TA) provided by the ADB. The NPRR proposes to amend ARIPO and provides mitigation measures for all those who are involuntary displaced, that is, those displaced by development projects and persons affected by river erosion and evictions displaces (non-projects).

Alleviation measures identified in the policy are comprehensive as they cover all the DP guidelines on resettlement. The draft policy proposed the same compensation package to those affected by the acquisition and requisition of land for development projects as well as for non-project (slum eviction and river erosion) affected. Provision for an identical compensation package has been made in the draft NPRR (Khatun 2015) for those impacted by (a) public sector projects, irrespective of the source of funding,(b) private sector and PPP projects, and (c) river erosion and eviction displaces. The draft NPRR mandates appropriate assistance and rehabilitation with special attention to the vulnerable that are unable to absorb the risks and cost of modern development (Halcrow et al. 2007).

Further, the draft policy ensures appropriate assistance to affected people and communities to restore and improve their socio-economic conditions and establish community social systems and networks during and after resettlement has taken place. It is expected that overall socio-economic development, the objective of any development project or nation as whole, can be achieved with the appropriate implementation of draft NPRR of 2007.

However, the draft policy needs to be endorsed at the political level and get the legal status to amend the ARIPO to provide for a comprehensive R&R package, which will strengthen the government's legal framework and effectively mitigate the 'impoverishment risks of involuntary displacement'.

The author recommends the following measure to improve the R&R status in Bangladesh and restore or improve livelihood of the APs:

(a) Resettlement implementation should commence prior to the civil work for the project
(b) Adequate time needs to be given to the AP to relocate after receiving the compensation
(c) Besides, cash compensation, mitigation measures for the restoration of livelihoods such as skill training, credit facilities and counselling should be arranged for
(d) Special medical facilities to children, women and disabled should be provided
(e) APs and their family members should be encouraged to explore various employment avenues in order to restore their livelihoods
(f) Community-based economic activities need to be encouraged and facilitated though the development project

(g) While developing the resettlement sites, special attention needs to be given to spatial arrangements in order to facilitate community cohesion and its sustenance
(h) A careful consideration of the geographical and cultural context in which the RPs is being implemented is essentials. Guidelines provided by DPs are of a general nature and may not be appropriate for a country or a specific area within a country
(i) The GoB need to take adequate measures to control encroachment on government land
(j) An unequal compensation for the DP funded projects creates hurdles at the implementation level due to dissatisfaction among APs of government-aided projects where the entitlements are much less. Development is for everyone and, therefore, the GoB should endorse and give legal status to the draft NPRR of 2007 that provides for identical reparation for development induced displacements and for non-project induced impacts such as river erosion and slum evictions. An equal compensation package to all the victims of involuntary displacement would ensure effective reconstruction of livelihood and reduce poverty.

References

Asian Development Bank (ADB). 1999. *Special Study on the Policy Impact of Involuntary Resettlement, Bangladesh Case Studies, Jamuna Multipurpose Bridge Project*. Manila: Asian Development Bank.

Asian Development Bank (ADB). 2003a. *Evaluation of Resettlement Experience in Selected Projects*. Draft Final Reports (II). Dhaka: Asian Development Bank and Roads and Highways Department.

Asian Development Bank (ADB). 2003b. 'Enhancing Capacity of Infrastructure Agencies in Management of Involuntary Resettlement Manila'. In *Enhancing Capacity of Infrastructure Agencies in Management of Involuntary Resettlement*. Dhaka: Asian Development Bank and Roads and Highways Department.

Cernea, M. M. 1996. 'Understanding and Preventing Impoverishment From Displacement Reflections on the State of Knowledge'. In M. Cernea and C. McDowell (eds.), *Understanding Impoverishment: The Consequences of Development-induced Displacement*, pp. 13–32. Oxford: Berghahn Books.

Cernea, M. M. 1998. 'Involuntary Resettlement in Development Projects: Bank-assisted Projects: Policy Guidelines in the World Bank-assisted Project'. *World Bank Technical* (80). Washington, DC: The World Bank.

Cernea, M. M. 2000. 'Risks, Safeguards and Reconstruction: A Model for Population Displacement and Resettlement'. In M. M. Cernea and C. McDowell (eds.), *Risks and Reconstruction: Experiences of Resettlers and Refugees*, pp. 11–55. Washington, DC: The World Bank.

Cernea, M. M. 2008. 'Reforming the Foundation of Involuntary Resettlement: Introduction'. In M. M. Cernea and H. M. Mathur (eds.), *Can Compensation Prevent Impoverishment? Reforming Resettlement Through Investment and Benefit Sharing*, pp. 1–10. Oxford: Oxford University Press.

Halcrow Group Limited, Bangladesh Consultants Limited and Rural Management Consultants Limited. 2007. *Development of a National Policy on Involuntary Resettlement*. Government of Bangladesh, ADB and Ministry of Land.

Khatun, H. 2000a. *Development-induced Displacement and Resettlement Practice in Bangladesh*. Paper presented at the 10th World Congress of Rural Sociology; Involuntary Resettlement: Risks Reconstruction and Development, Rio de Janeiro, Brazil, 30 July–5 August.

Khatun, H. 2000b. *Resettlement Process in Jamuna Multipurpose Bridge Project: An Induced Displacement and its Mitigation*. International Conference on Disaster: Issues and Gender Perspectives, Bangladesh Geographical Society, Dhaka, 23 and 24 June.

Khatun, H. 2005. 'Development-induced Displacement and Rehabilitation in Bangladesh: Challenges and Perspectives'. *Oriental Geographer*, 49(2): 135–150.

Khatun, H. 2009. 'Displacement and Poverty: Measures for Restoring Meagre Livelihoods'. In R. Modi (ed.), *Beyond Relocation: The Imperative of Sustainable Resettlement*, pp. 331–353. New Delhi, India: SAGE Publications.

Khatun, H. 2015. *Role of Participatory Supervision in Restoration of Livelihood of APs: Case Study of Bhairab Bridge Project, Bangladesh*. Presented in IDEAS Global Assembly Evaluating Sustainable Bangkok, Thailand, 26–30 October 2015.

Wet, de C. 2004. 'Contested Spaces: An Outside Perspective on Database on Development-Induced Displacement and Resettlement (DIDR) in India'. In *International Seminar on Development and Displacement: Afro-Asian Perspective*. Hyderabad: Osmania University, India.

Zaman, M. Q. 1996. 'Development and Displacement in Bangladesh: Towards a Resettlement Policy'. *Asian Survey*: 691–703.

12 Geographical methods in undertaking disaster and vulnerability research

Shitangsu Kumar Paul

1. Introduction

Scholars from various disciplines use different meanings and concepts of vulnerability, which has led them to adopt different methodological approaches to assess it. Differences in such methodological approaches to assess vulnerability among the disciplines can be elucidated by their propensity to focus on different elements of risk, responses to such risk and outcomes. The objective of this chapter is to review various models and present geographical methods of understanding vulnerability in disaster research. In disaster discourse, vulnerability is deeply rooted in several components, such as risk or risky events, alternatives for managing risk or responses to the risk and outcomes in terms of minimizing loss. However, most of the disciplines put emphasis either on the risks on the one hand or the underlying conditions on the other. In view of the fact that each discipline has its own rationale for defining and assessing vulnerability, there is hence no basis to consider that concepts, measures and methods will be similar across the disciplines. Therefore, the chapter reviewed various models to understand vulnerability from a natural hazard perspective. The chapter suggests for the preparation of separate vulnerability maps for each hazard. Moreover, it emphasizes using this approach on vulnerability rather than risk in an attempt to analyze physical and social factors in the implementation of natural hazard risk assessment.

2. Vulnerability definitions and relevant terminologies

It is true that vulnerability has no universal definition, but undoubtedly it is a powerful analytical tool in describing the existing condition of susceptibility to harm, powerlessness and marginality of both physical and socio-ecological systems. At the same time, for guiding normative

Table 12.1 Selected definitions of vulnerability in disaster discourse

Author(s)	Definitions
Gabor and Griffith (1980)	Vulnerability is the threat (of hazardous materials) to which people are exposed (including chemical agents and the ecological situation of the communities and their level of emergency preparedness). Vulnerability is the risk context
Timmerman (1981)	Vulnerability is the degree to which a system acts adversely to the occurrence of a hazardous event. The degree and quality of the adverse reaction are conditioned by a system's resilience (a measure of the system's capacity to absorb and recover from the event)
UNDRO (1982)	Vulnerability is the degree of loss to a given element or set of elements at risk resulting from the occurrence of a natural phenomenon of a given magnitude
Susman et al. (1984)	Vulnerability is the degree to which different classes of society are differentially at risk
Kates (1985)	Vulnerability is the 'capacity to suffer harm and react adversely'
Pijawka and Radwan (1985)	Vulnerability is the threat or interaction between risk and preparedness. It is the degree to which hazardous materials threaten a particular population (risk) and the capacity of the community to reduce the risk or adverse consequences of hazardous materials releases
Bogard (1988)	Vulnerability is operationally defined as the inability to take effective measures to insure against losses. When applied to individuals, vulnerability is a consequence of the impossibility or improbability of effective mitigation and is a function of our ability to select the hazards
Mitchell (1989)	Vulnerability is the potential for loss
Liverman (1990)	Distinguishes between vulnerability as a biophysical condition and vulnerability as defined by political, social and economic conditions of society. She argues for vulnerability in geographical space (where vulnerable people and places are located) and vulnerability in social space (who in that place is vulnerable)
Downing (1991)	Vulnerability has three connotations: it refers to a consequence (e.g., famine) rather than a cause (e.g., are vulnerable to hunger); and it is a relative term that differentiates among socio-economic groups or regions, rather than an absolute measure of deprivation
Dow (1992)	Vulnerability is the differential capacity of groups and individuals to deal with hazards, based on their positions within physical and social worlds

(*Continued*)

Table 12.1 (Continued)

Author(s)	Definitions
Smith (1992)	Risk from a specific hazard varies through time and according to changes in either (or both) physical exposure or human vulnerability (the breadth of social and economic tolerance available at the same site)
Alexander (1993)	Human vulnerability is a function of the costs and benefits of inhabited areas at risk from natural disasters
Cutter (1993)	Vulnerability is the likelihood that an individual or group will be exposed to and adversely affected by a hazard. It is the interaction of the hazards of place (risk and mitigation) with the social profile of communities
Watts and Bohle (1993)	Vulnerability is defined in terms of exposure, capacity and potentiality. Accordingly, the prescriptive and normative response to vulnerability is to reduce exposure, enhance coping capacity, strengthen recovery potential and bolster damage control (i.e., minimize destructive consequences) via private and public means
Blaikie et al. (1994)	By vulnerability we mean the characteristics of a person or group in terms of their capacity to anticipate, cope with, resist and recover from the impact of a natural hazard. It involves a combination of factors that determine the degree to which someone's life and livelihood is put at risk by a discrete and identifiable event in nature or in society
Bohle et al. (1994)	Vulnerability is best defined as an aggregate measure of human welfare that integrates environmental, social, economic and political exposure to a range of potential harmful perturbations. Vulnerability is a multi-layered and multidimensional social space defined by the determinate, political, economic and institutional capabilities of people in specific places at specific times
Dow and Downing (1995)	Vulnerability is the differential susceptibility of circumstances contributing to vulnerability. Biophysical, demographic, economic, social and technological factors such as population ages, economic dependency, racism and age of infrastructure are some factors which have been examined in association with natural hazards
Gilard and Givone (1997)	Vulnerability represents the sensitivity of land use to the hazard phenomenon
Comfort et al. (1999)	Vulnerability is those circumstances that place people at risk while reducing their means of response or denying them available protection

Table 12.1 (Continued)

Author(s)	Definitions
Weichselgartner and Bertens (2000)	By vulnerability we mean the condition of a given area with respect to hazard, exposure, preparedness, prevention and response characteristics to cope with specific natural hazards. It is a measure of capability of this set of elements to withstand events of a certain physical character
Adger (2006)	Vulnerability is the state of susceptibility to harm from exposure to stresses associated with environmental and social change and from the absence of capacity to adapt
Ciurean et al. (2013)	Vulnerability refers to the inability to withstand the effects of a hostile environment
Wolf et al. (2013)	Vulnerability, in ordinary language, is a measure of possible future harm
Proag (2014)	Vulnerability has been defined as the degree to which a system, or part of a system, may react adversely during the occurrence of a hazardous event
Mechler and Bouwer (2015)	Vulnerability to natural hazards, defined as the propensity to incur losses in a hazardous event
Chakraborty and Joshi (2016)	Vulnerability integrates social and environmental systems to reduce the intensity and frequency of risks
Pazzi et al. (2016)	Vulnerability is defined as the degree of loss of a given set of elements at risk resulting from the occurrence of a natural phenomenon of a certain magnitude
Ferraro and Schafer (2017)	Vulnerability is a lack of resources in one or more life domains, which, given a specific context, places individuals or groups at a major risk of experiencing negative consequences related to sources of stress, the inability to cope effectively with stressors and the inability to recover from the stressor or to take advantage of opportunities before a given deadline

Source: Adapted from Cutter (1996), Weichselgartner (2001), Hogan and Marandola (2005), Paul (2013)

analysis of measures to enhance well-being through the reduction of disaster risk (Adger 2006). However, for a clear understanding, a few selected definitions of vulnerability are presented in Table 12.1.

3. Vulnerability, hazard and disaster: review of models

A number of research works were done at the beginning of the 1940s regarding human occupancy in hazardous places, adjustment to minimize impacts and the social acceptance or at least tolerance of the risks associated with placing human lives and livelihoods in harmful way. White et al. (1958) derived the risk/hazards paradigm to understand who lived in hazardous areas and the drivers of the nation's increasing vulnerability to losses from natural hazards (Burton et al. 1978). Such natural event exposure-based approach sustained for three decades until researchers began to question the rationality of such a natural event centric approach. Since then, most of the studies on human response to natural hazard and the use of socio-economic variables responsible for vulnerability to hazardous events followed two different discourses. In a social stratification approach, some find no relationships while others come across positive associations. Therefore, disagreement exists on which socio-economic variables should be considered. The traditional views follow a social stratification approach and use distinct groupings of relevant variables, while other approaches reject the appropriateness of using the cultural indicators of social stratification as true indicators of socio-economic association. The latter approach supports the use of social class analysis to explain variation in human vulnerability to hazards (Paul and Routray 2010,2011).

In the 1990s, natural hazards scholars started to focus on the issue of vulnerability of people owing to the impacts of environmental change, especially climate change. In this case geography provides the major disciplinary legacy (Paul 2013), as it covers geographic space (where vulnerable people and places are located) or social space (who in those places are most vulnerable), with an application of scales ranging from local, regional to global. Human geography and human ecology have theorized vulnerability to environmental change and have made contributions to present understanding of social-ecological systems, while related insights into entitlements grounded vulnerability analysis theories put emphasis on social change and decision-making fields (Adger 2006). Moreover, vulnerability research in the hazards tradition is defined by three overlapping areas, i.e., human ecology or political ecology, natural hazards and the so-called 'Pressure and Release' model that covers the space between hazards and political ecology approaches (Adger 2006). A few models related to vulnerability and disasters are reviewed and presented in what follows:

3.1 The double structure of vulnerability

According to the double structure of vulnerability model by Bohle (2001) the upper part is an external side associated with the exposure to risks and shocks. This part is controlled by a political economy approach, such as social discriminations, uneven distribution of resources; human ecology, for example, population dynamics and environmental management capacities; and the rights-based approach, such as the inability to obtain resource through proper economic means. Another part of the model is the internal side related to the coping capacity to expect, cope with, resist and recover from the shock of a hazard and is inclined to the Crisis and Conflict Theory (control of assets and resources, capacities to manage crisis situations and resolve conflicts), Action Theory, such as the free reaction of people towards social, economic or governmental constrains and how vulnerability can be mitigated through access to assets. Ciurean et al. (2013) agree that the theoretical model of the double structure of vulnerability cannot effectively be considered without considering coping and human response capability.

3.2 Risk and hazard model

The Risk-Hazard (RH) model aims 'to understand the impact of hazard as a function of exposure to the hazard event and the dose-response (sensitivity) of the entity exposed' (Turner et al. 2003). Professionals dealing with disaster risks considered vulnerability as a component within the context of hazard and risk. Vulnerability, coping capacity and exposure are viewed as separate features. Davidson's conceptual framework views risk as the sum of hazard, exposure, vulnerability and capacity measures. Davidson's (1997) conceptual model was later modified by Bollin et al. (2003). Hazard is characterized by probability and severity; exposure is characterized by structure, population and economy; however, vulnerability is characterized by physical, social, economic and environmental dimension, capacity and measures are associated with physical planning and management, as well as social and economic ability.

3.3 Vulnerability in the global environmental change

Turner et al. (2003) introduced another conceptual model on the basis of the global environmental change, mainly focusing on the coupled human-environment systems. The model encircles vulnerability as exposure, sensitivity and resilience. Exposure is defined as a set of components, such as threatened elements – individuals, households, states, ecosystem, etc. that are subjected to damage and characteristics of threat such as frequency, magnitude and duration. The sensitivity is determined by human factors, such as social capital and endowments and environmental conditions, such as natural capital or biophysical endowments by which the resilience is influenced and adjustment and adaptation enhance the last component. In addition, vulnerability to hazards of a system consists of linkages to the broader human and environmental conditions and processes operating on the coupled system in question; perturbations and stressors/stresses (Farrell et al. 2001) – that emerge from this conditions and processes; and the coupled human – the environment system of concern in which vulnerability resides, including exposure and responses (i.e., coping, impacts, adjustments and adaptation) (Turner et al.2003).

3.4 The Pressure and release model

The Pressure and Release (PAR) model explicitly defines 'risk as a function of the perturbation, stressor or stress and the vulnerability of the exposed unit' (Turner et al. 2003). The model works at different spatial scales, such as place, region and world, etc., including functional and temporal scales that analyze the interaction of the multiple perturbations and stressor/stresses (Wisner et al. 2004). Hazards are considered as being controlled by the inside and outside of the analyzed system but they are commonly considered site-specific according to their character. As a result, there raises complexity in place of assessment is the main factor to locate hazards. The model is formulated on the normally used equation, which describes risk as a function of the hazard and vulnerability. It highlights the original driving forces of vulnerability and the conditions present in a system that contribute to disaster situations when a hazard occurs. Vulnerability is linked with these conditions at three progressive levels, for example:

(a) *Root causes*, which can be defined as limited access to power structures or resources; another way it can be related with political ideologies or economic systems
(b) *Dynamic pressures*, which can be represented by demographic or social changes in time and space such as rapid population decrease, rapid urbanization, lack of local institutions, appropriate skills or training, etc.
(c) *Unsafe conditions* posed by the physical environment, such as unprotected buildings and infrastructure, dangerous slopes, etc., other way it can be a socio-economic context, such as lack of local institutions, prevalence of endemic diseases, etc. Birkmann (2006) argues that such a conceptual framework is a significant approach that goes beyond the identification of vulnerability towards dealing with its root causes and the dynamic forces rooted in the human-environment structure

A few limitations of the model are that it does not address the coupled human-environment system in the sense of considering the vulnerability of biophysical subsystems (Kasperson et al. 2003); it provides little detail on the structure of the hazard's causal sequence, including the nested scales of interactions; and it tends to underemphasize feedback beyond the system of analysis that integrative RH models include (Davidson 1997).

3.5 A holistic approach to risk and vulnerability

Vulnerability is considered by three categories of factors in the holistic approach to risk and vulnerability (Carreño et al. 2007). These categories include: physical exposure and susceptibility – regarded as hazard dependent; fragility of the socio-economic system – non-hazard dependent; and lack of resilience to cope and recover – non-hazard dependent. In this approach, comprehensive and multidisciplinary perspectives were given and emphasized for measuring vulnerability. This model considers direct physical impacts such as exposure and susceptibility and at the same time indirect consequences like socio-economic fragility and lack of resilience to a possible hazardous event. In each category, sets of indicators or indices explain the vulnerability factors. The model consists of a control system that changes indirectly the level of risk in corrective and prospective ways. Within each category, the vulnerability factors are explained with sets of indicators or indices. According to Ciurean et al. (2013), the model includes a control system which alters indirectly the level of risk through corrective and prospective intrusions (risk identification, risk reduction, disaster management).

3.6 Disaster Resilience of Place (DROP) model

Cutter et al. (2008) developed a conceptual framework called the Disaster Resilience of Place (DROP) model to address resilience to natural disaster and also present the relationship between resilience and vulnerability. The model is place-based and focuses on social resilience of a certain place. This model was developed for natural hazards contexts, however it can also be used in other similar contexts, such as climate change related risks. The conceptual development occurred with the hazard-of-place approach to vulnerability, that is, a hybrid of the risk/hazard and political ecology perspectives. First formulated in 1996, it describes the place-based interaction between biophysical vulnerability (exposure) and social vulnerability in an overall determination of the differential social burdens of hazards and how this relationship changes over time and across space (Cutter 1996). The model is built on the antecedent conditions of the system, meaning the vulnerability and the resilience of the system dependent upon interaction with the natural systems and the social system and built environment.

The main drawbacks of the approach are its weaknesses in identifying the root causes of social vulnerability and the failure to take into account the bigger context where such vulnerability exists. There are number of indicators that can be used to understand community resiliency to natural hazards. Cutter et al. (2008) argue that there are so far no successful efforts in forming a model to measure or model resilience at local level. However, even though this model has a few weaknesses, the approach is most agreeable to empirical testing and the use of geospatial techniques.

However, different scope and thematic focuses are illustrated in the previously discussed models. The vulnerability definition is explained through exposure, coping capacities, sensitivity and adaptation response in the model of double structure of vulnerability (Bohle 2001) and, in the global environmental change model (Turner et al. 2003), vulnerability is different from these characteristics in consideration of the framework of hazard and risk approach. Factors and conditions of vulnerability indicated in the holistic approach and PAR model are able to measure direct physical impacts and at same time indirect effects of disasters. Besides, the DROP model presents the relationship between resilience and vulnerability. It is evident that there is no universal model that can specify all specific needs but a specific approach serves for certain disciplinary groups' needs. We can understand that vulnerability is explained by these conceptual approaches. Some approaches are used as models of quantitative vulnerability assessment and some are for qualitative vulnerability assessment.

4. Common geographical method of vulnerability research: a multi-criteria approach

Weichselgartner (2001) used qualitative methods to measure hazard, exposure, preparedness, prevention and response to a specific hazard. Following a disaster management cycle, a similar method was applied by Dewan (2013) using Geographical Information Systems (GIS) as a remote sensing tool for disaster management. However, Weichselgartner (2001) used mappable indicators to assess each factor according to the scale of analysis. Vulnerability to specific hazards is assessed by the overlay of each theme. In this regard, a multi-hazard approach is also possible; nonetheless, it is preferable to prepare each single map on specific hazards and an overlay of different thematic maps to assess vulnerability (Preston et al. 2011). A Geographical Information System (GIS) can play a vital role invisualization and spatial analysis of vulnerability mapping. The potential benefits of vulnerability mapping are twofold. First, they help support spatial planning (Clark et al.1998; NRC 2007). Second, vulnerability mapping has a role to play in instructing the public about disaster and climate change and the mechanisms by which it may interact with coupled human/environmental systems (Preston et al. 2009).

According to Weichselgartner (2001), vulnerability usually puts emphasis on biophysical, technological hazards and coping responses to withstand against such hazards. Recently, Weichselgartner and Bertens (2000) proposed a combined form of both of these elements and considered vulnerability as both a biophysical hazard and social response together in a specific geographic domain. A vulnerability map of a specific hazard on a given area in a specific time with a specific scale can be prepared after successfully overlaying each successive thematic map. However, it is preferable to prepare separate maps on specific themes on a specific hazard, and subsequently various vulnerability maps can be overlaid and interpreted to get a final vulnerability map. Following successive steps is involved to identify the vulnerability of an area.

4.1 Natural hazard analysis

In a disaster management cycle, the first step is the analysis of the natural hazard. In this stage, identification, inventory and assessment of all natural events in a given place which can potentially disrupt human life and property are identified. A hazard map is prepared on the basis of a physical process related to the studied natural event. This map qualitatively and quantitatively represents a hazard event, such as characteristics and process of occurring, probability of occurrence and intensity. According to Weichselgartner (2001), major factors related to the hazard event are magnitude, intensity, frequency, duration and destructiveness, speed of onset, distribution and predictability of processes.

4.2 Exposure analysis

Exposure analysis includes identification, inventory and assessment of infrastructure, property and individuals, etc. in a specific area. Besides, direct and indirect impacts of hazard event are also considered. Social structure and infrastructure related variables are analyzed in order to prepare the exposure map. The prepared map portrays areal extent of hazard event, social structure and infrastructure related variables, which have possibilities of being affected. According to Weichselgartner (2001), the most important factors of exposure analysis are susceptibility of building contents to damage, robustness of building fabric, key installations and public supply services, transportation systems, population distribution and density and land use activity.

4.3 Preparedness analysis

The preparedness analysis includes identification, inventory and assessment of preparedness measures undertaken in a given area to combat the impending disasters. Disaster warning, evacuation and relief variables are included in the analysis to prepare a preparedness map in order to present precautionary preparedness activities and measures. According to Weichselgartner (2001), the indicator used for this map is a binary form to characterize factors such as awareness, warning and evacuation.

4.4 Prevention analysis

The prevention analysis includes identification, inventory and assessment of all strategies in a specific area to prevent and minimize hazards and their impacts. Analyses of structural and non-structural measures are needed to generate prevention map. Weichselgartner (2001) suggested including various factors, for example water control measures, land use and infrastructure control measures, as well as financial relief and loss reduction measures to prepare prevention map.

4.5 Response analysis

Response analysis includes identification, inventory and assessment of all response activities in a specific area to minimize social and economic losses. Search, rescue, humanitarian assistance, recovery and reconstruction analysis are needed to prepare the response map. Important factors that need to be considered are emergency oriented and response related, for example, rescue, relief, humanitarian assistance, etc.

4.6 Vulnerability analysis

Vulnerability analysis includes assessment of the existing condition of a specific area and capacity to withstand to a specific hazard event.

It includes the analysis of the hazard event or hazard characteristics, socio-economic condition, exposure, preparedness, prevention and response variables. A separate map for each theme is generated in their respective sections. In this regard, a vulnerability map can be generated through the overlay of each thematic map. An average class value can be obtained assuming that all features have equal weight. In this regard, previous maps, such as hazard, socio economic exposure, preparedness, prevention and response maps are used for generating the vulnerability map.

5. Multi-criteria analysis for vulnerability assessment and the role of geography

Spatial Multi-Criteria Analysis (MCA) or Multi-Criteria Evaluation (MCE) has received attention because it provides improved decision-making (Malczewski 2004, 2006); develops and evaluates alternative plans (Malczewski 1996); and is primarily suitable for spatial decision-making, as the data that the decision-makers rely on is mostly related to space (Malczewski 1999). Even though the recent development of spatial information systems permits the integration of diverse data with MCA/MCE methods, some researchers have employed this technique for water resource management, particularly flood risk assessment and disaster management (Meyer et al. 2009). Newton and Weichselgartner (2014) argue that coastal societies can be vulnerable to multiple threats, so an integrated multi-criteria approach is necessary to determine coastal vulnerability. A multi-criteria index was applied using a multi-layer hierarchical model to classify vulnerability, considering three dimensions such as flood exposure, socio-economic sensitivity and infrastructure in the Amazon delta (Yorke et al. 2015; Mansur et al. 2016) used geographic information system to model the infrastructure, context analysis, geophysical risk analysis and social vulnerability analysis as a component for assessing the overall place vulnerability. A GIS-based spatial multi-criteria model is used to assess the spatial vulnerability to earthquake of Bucharest, Romania. Environmental, socio-economic and physical dimensions are included in the semi-quantitative Analytic Hierarchy Process (AHP) to identify vulnerability at the individual sector and to identify a 'composite' index map classified in various qualitative classes, comparing the histogram from the overall vulnerability (Armas 2012). Armas and Gavris (2013) further used the Social Vulnerability Index (SoVI) and the Spatial Multi-Criteria Social Vulnerability Index (SEVI) model to aggregate complex indicators to assess social vulnerability in Bucharest, Romania. Authors found that after applying SoVI and SEVI, the results are not similar and they concluded that both the methods cannot be used interchangeably.

A multi-hazard zoning map of Bangladesh is prepared, considering cyclone, tornado, flood and earthquake. In this regard, a historical database of previously mentioned hazards, intensity scales and damage risk levels were developed. Based on this, district-wise individual and multi-hazard scores were calculated and finally prepared a district-wise multi-hazard zoning map of Bangladesh, divided into high, moderate and low hazard zones (Barua et al. 2016). Similarly, Hazarika et al. (2018) assessed hazard, vulnerability and risk due to flooding using an indicator-based methodology, incorporating stakeholder knowledge and multi-criteria evaluation in geographic information system to achieve community-based assessment in the upper Brahmaputra river valley. Results show spatial distribution of hotspots of flood hazard and vulnerability and locations at risk at regional and sub-regional level. The emergent risk pattern indicates that vulnerability indicators have more significant contribution than hazard indicators in the upper Brahmaputra valley. Likewise, Ahsan and Warner (2014) developed a socio-economic vulnerability index for climate change induced disaster affected communities in south-western Bangladesh. Authors individually identified demographic, social, economic and physical vulnerability; and, finally, measured overall vulnerability index of each individual area. Results show that southern and south-eastern unions are relatively more vulnerable to natural hazards, although these unions are surrounded by mangroves. Besides, socio-economic condition and disaster frequency are identified as crucial factors of such vulnerability, adaptive capacity, sensitivity and exposure of the region. Likewise, Rahman et al. (2015) assessed earthquake and fire hazard vulnerability using the visual screening method FEMA-RVS for earthquake, the ADPC method for fire hazard and the World Bank method for social vulnerability. Physical and social vulnerability scores are combined together by using the method given by Cardona (2005). Results shows that 11.7 percent are very highly vulnerable, 26.3 percent are highly vulnerable and 42 percent are moderately vulnerable to both fire and earthquake. Besides, the studied area is relatively more vulnerable to fire hazard than earthquake because of dense population and other socio-economic factors. A Coastal Vulnerability Index (CVI) was developed by Islam et al. (2016) using geospatial tools including various physical variables such as geomorphology, coastal slope, shoreline change rate, sea level change rate, mean tide range, coastal bathymetry and storm surge height. Results show that 37.6, 21.5 and 40.9 percent areas are very high, moderate and low vulnerable, respectively. It is noted that the rate of erosion and tidal range are important factors at the extreme edge of the Ganges delta while rate of accretion and sea level rise appear to play a minor role in the inner part. Likewise, Ahmed (2015) applied Multi-criteria Decision Analysis, Artificial Hierarchy Process (AHP), Weighted Liner Combination (WLC) and Ordered Weighted Average (OWA) methods in the Chittagong Metropolitan Area to identify landslide susceptibility and found satisfactory agreement between the produced susceptibility map and historical data on 20 critical landslide locations. Alam et al. (2017) used Livelihood Vulnerability Index (LVI) and Climate Vulnerability Index among resource poor char and riverbank erosion prone areas of Bangladesh and found households inhabiting in the char areas are more vulnerable than riverbank erosion areas. Identified drivers are livelihood strategies and access to food, water and health facilities. On the other hand, erosion prone inhabitants are vulnerable because of river morphology driven erosion, loss of land and decrease in economic potentials.

Social vulnerability in the context of cyclone Sidr and Aila is also analyzed by various authors for sustainable coastal belt development planning in Bangladesh (Mallick et al. 2011; Paul and Routray 2011; Hossain 2015; Paul 2015). Authors found that cyclone impact varies among individuals, groups and communities; and the affected community rely on relief materials, which creates relief dependency and makes them more vulnerable to impending disasters; in the long run, they fall in a 'vulnerability trap'. Likewise, households having dilapidated housing conditions, lower levels of income and education are less likely to receive weather forecasting and disaster training and are more likely to become vulnerable to disasters (Paul and Routray 2010,2011; Hossain 2015; Paul 2015). A few authors have identified the complex relationship between environmental risk, poverty and vulnerability in the south-east Bangladesh, considering household and community vulnerability and coping mechanisms, and found that households with lower income and limited access to productive assets face higher exposure to flooding events. Besides, income and asset disparity trend to higher exposure level (Brouwer et al.2007; Paul 2012, 2013,2015).

From the previous discussions it can be said that geography as a discipline is particularly appropriate for conducting a combined and interdisciplinary vulnerability research due to its exceptional capacity for problem solving, considering the place, space and scale of analysis. Such capacity is reasonable because of subject matter and scope of the discipline. In addition, rapid expansion of the power and accessibility of geospatial tools and data has enhanced the potentials for creating a common platform (i.e., multi-criteria analysis) for multidisciplinary and integrated vulnerability research (Paul 2013).

6. Concluding remarks

Various definitions of vulnerability across the disciplines present different methodological approaches to measure vulnerability and their propensity to focus on various elements of risk, response to such risk and outcomes. The vulnerability discourse faces the problem of validity and constructing in an integrated, universally agreed upon research framework. In this regard, geography lies in the society-nature interface. Hence, it can contribute to widening the dialogue among science in general about the construction of a new paradigm, as well as ontological and epistemological methods for operationalizing vulnerability science. Progress has been done in this field using empirical as well as statistical methods aided by a Geographic Information System and advanced computational models (MCA/MCE/SoVI/SEVI/AHP) are used to estimate uncertainty in vulnerability and its components. In this regard, Weichselgartner's (2001) multi-criteria approach to measure vulnerability to disaster is easily applicable, because it uses a series of indicators combined in a simple way with clearly defined steps, and the final outcome is the identification of vulnerability. This approach creates vulnerability maps using binary data. Such maps not only present degrees of vulnerability but also the causes of vulnerability. Moreover, the combination of various vulnerability maps (i.e., hazard, exposure, preparedness, prevention and response) is possible; however, the interpretation of such maps would be difficult. Therefore, preparation of separate vulnerability maps for each hazard is suggested. Moreover, this approach puts emphasis on vulnerability rather than on risk in an attempt to analyze physical and social factors in the implementation of natural hazard risk assessment. However, using such an approach for detailed micro/local level vulnerability assessment is easier with a regional application than macro-level.

References

Adger, W. N. 2006. 'Vulnerability'. *Global Environmental Change*, 16(3): 268–281.
Ahmed, B. 2015. 'Landslide Susceptibility Mapping Using Multi-Criteria Evaluation Techniques in Chittagong Metropolitan Area, Bangladesh'. *Landslides*, 12(6): 1077–1095.

Ahsan, M. N. and J. Warner. 2014. 'The Socioeconomic Vulnerability Index: A Pragmatic Approach for Assessing Climate Change Led Risks: A Case Study in the South-Western Coastal Bangladesh'. *International Journal of Disaster Risk Reduction*, 8: 32–49.

Alam, G. M., K. Alam, S. Mushtaq and M. L. Clarke. 2017. 'Vulnerability to Climatic Change in Riparian Char and River-Bank Households in Bangladesh: Implication for Policy, Livelihoods and Social Development'. *Ecological Indicators*, 72: 23–32.

Alexander, D. 1993. *Natural Disasters*. New York: Chapman and Hall.

Armaş, I. 2012. 'Multi-Criteria Vulnerability Analysis to Earthquake Hazard of Bucharest, Romania'. *Natural Hazards*, 63(2): 1129–1156.

Armaş, I. and A. Gavriş. 2013. 'Social Vulnerability Assessment Using Spatial Multi-criteria Analysis (SEVI Model) and the Social Vulnerability Index (SoVI Model) – A Case Study for Bucharest, Romania'. *Natural Hazards and Earth System Sciences*, 13(6): 1481–1499.

Barua, U., M. S. Akhter and M. A. Ansary. 2016. 'District-wise Multi-Hazard Zoning of Bangladesh'. *Natural Hazards*, 82(3): 1895–1918.

Birkmann, J. 2006. 'Measuring Vulnerability to Promote Disaster-resilient Societies: Conceptual Frameworks and Definitions'. In J. Birkmann (eds.), *Measuring Vulnerability to Natural Hazards: Towards Disaster Resilient Societies*, pp. 9–54. Tokyo: United Nations University Press.

Blaikie, P., T. Cannon, I. Davis and B. Wisner. 1994. *At Risk: Natural Hazards, People's Vulnerability, and Disasters*. London: Routledge.

Bogard, W. C. 1988. 'Bringing Social Theory to Hazards Research: Conditions and Consequences of the Mitigation of Environmental Hazards'. *Sociological Perspectives*, 31: 147–168.

Bohle, H. G. 2001. 'Vulnerability and Criticality: Perspectives From Social Geography'. *IHDP Update*, 2(1): 3–5.

Bohle, H. G., T. E. Downing and M. J. Watts. 1994. 'Climate Change and Social Vulnerability: Toward a Sociology and Geography of Food Insecurity'. *Global Environmental Change*, 4(1): 37–48.

Bollin, C., C. Cardenas, H. Hahn and K. S. Vatsa. 2003. 'Natural Disaster Network: Disaster Risk Management by Communities and Local Governments'. In *Regional Policy Dialogue*. Washington, DC: Inter-American Development Bank.

Brouwer, R., S. Akter, L. Brander and E. Haque. 2007. 'Socioeconomic Vulnerability and Adaptation to Environmental Risk: A Case Study of Climate Change and Flooding in Bangladesh'. *Risk Analysis*, 27(2): 313–326.

Burton, I., R. W. Kates and G. F. White. 1978. *The Environment as Hazard*. New York: Guilford Press.

Cardona, O. D. 2005. *Indicators of Disaster Risk and Risk Management: Program for Latin America and the Caribbean: Summary Report*. Washington, DC: Inter-American Development Bank.

Carreño, M. L., O. D. Cardona and A. H. Barbat. 2007. 'A Disaster Risk Management Performance Index'. *Natural Hazards*, 41(1): 1–20.

Chakraborty, A. and P. K. Joshi. 2016. 'Mapping Disaster Vulnerability in India Using Analytical Hierarchy Processes'. *Geomatics, Natural Hazards and Risk*, 7(1): 308–325.
Ciurean, R. L., D. Schröter and T. Glade. 2013. 'Conceptual Frameworks of Vulnerability Assessments for Natural Disasters Reduction'. In J. Tiefenbacher (ed.), *Approaches to Disaster Management – Examining the Implications of Hazards, Emergencies and Disasters*, 1: 1–32.
Clark, George E., S. C. Moser, S. J. Ratick, K. Dow, W. B. Meyer, S. Emani, W. Jin, J. X. Kasperson, R. E. Kasperson and H. E. Schwarz. 1998. 'Assessing the Vulnerability of Coastal Communities to Extreme Storms: The Case of Revere, MA., USA'. *Mitigation and Adaptation Strategies for Global Change*, 3(1): 59–82.
Comfort, L., B. Wisner, S. Cutter, R. Pulwarty, K. Hewitt, A. Oliver-Smith, J. Wiener, M. Fordham, W. Peacock and F. Krimgold.1999. 'Reframing Disaster Policy: The Global Evolution of Vulnerable Communities'. *Global Environmental Change Part B: Environmental Hazards*, 1(1): 39–44.
Cutter, S. L. 1993. *Living With Risk*. London: Edward Arnold.
Cutter, S. L. 1996. 'Vulnerability to Environmental Hazards'. *Progress in Human Geography*, 20(4): 529–539.
Cutter, S. L., L. Barnes, M. Berry, C. Burton, E. Evans, E. Tate and J. Webb. 2008. 'A Place-based Model for Understanding Community Resilience to Natural Disasters'. *Global Environmental Change*,18(4): 598–606.
Davidson, R. 1997. *An Urban Earthquake Disaster Risk Index*. The John A. Blume Earthquake Engineering Center, Report (121), Department of Civil Engineering, Stanford University.
Dewan, A. 2013. *Floods in a Megacity: Geospatial Techniques in Assessing Hazards, Risk and Vulnerability*. Springer: Dordrecht.
Dow, K. 1992. 'Exploring Differences in Our Common Future(s): The Meaning of Vulnerability to Global Environmental Change'. *Geoforum*, 23(3): 417–436.
Dow, K. and T. E. Downing. 1995. *Vulnerability Research: Where Things Stand*. National Emergency Training Centre.
Downing, T. E. 1991. 'Vulnerability to Hunger in Africa: A Climate Change Perspective'. *Global Environmental Change*, 1(5): 365–380.
Farrell, A., S. D. Vandeveer and J. Jager. 2001. 'Environmental Assessments: Four Under-Appreciated Elements of Design'. *Global Environmental Change*, 11(4): 311–333.
Ferraro, K. F. and M. H. Schafer. 2017. 'Visions of the Life Course: Risks, Resources, and Vulnerability'. *Research in Human Development*, 14(1): 88–93.
Gabor, T. and T. K. Griffith. 1980. 'The Assessment of Community Vulnerability to Acute Hazardous Materials Incidents'. *Journal of Hazardous Materials*, 3(4): 323–333.
Gilard, O. and P. Givone. 1997. 'Flood Risk Management: New Concepts and Methods for Objective Negotiations'. In G. H. Leavesley, H. F. Lins, F. Nobilis, R. S. Parker, V. R. Schneider and F. H. M. van de Ven (eds.), *Destructive*

Water: Water-caused Natural Disasters, their Abatement and Control, pp. 145–155. Oxfordshire: IAHS Press.

Hazarika, N., D. Barman, A. K. Das, A. K. Sarma and S. B. Borah. 2018. 'Assessing and Mapping Flood Hazard, Vulnerability and Risk in the Upper Brahmaputra River Valley Using Stakeholders' Knowledge and Multi-Criteria Evaluation (MCE)'. *Journal of Flood Risk Management*, 11: S700–S716.

Hogan, D. J. and E. Marandola.2005. 'Towards an Interdisciplinary Conceptualisation of Vulnerability'. *Population, Space and Place*, 11(6): 455–471.

Hossain, M. N. 2015. 'Analysis of Human Vulnerability to Cyclones and Storm Surges Based on Influencing Physical and Socioeconomic Factors: Evidences From Coastal Bangladesh'. *International Journal of Disaster Risk Reduction*, 13: 66–75.

Islam, M. A., D. Mitra, A. Dewan and S. H. Akhter. 2016. 'Coastal Multihazard Vulnerability Assessment Along the Ganges Deltaic Coast of Bangladesh: A Geospatial Approach'. *Ocean & Coastal Management*, 127: 1–15.

Kasperson, J. X., R. E. Kasperson, B. L. Turner, W. Hsieh and A. Schiller. 2003. 'Vulnerability to Global Environmental Change'. In A. Diekman, T. Dietz, C. C. Jaeger and E. A. Rosa (eds.), *The Human Dimensions of Global Environmental Change*. Cambridge, MA: MIT Press.

Kates, R. W. 1985. 'The Interaction of Climate and Society'. In R. W. Kates, H. Ausubel and M. Berberian (eds.), *Climate Impact Assessment*, Chapter 1. Chichester: Wiley.

Liverman, D. M. 1990. 'Drought Impacts in Mexico: Climate, Agriculture, Technology, and Land Tenure in Sonora and Puebla'. *Annals of the Association of American Geographers*, 80(1): 49–72.

Malczewski, J. 1996. 'A GIS-based Approach to Multiple Criteria Group Decision-Making'. *International Journal of Geographical Information Systems*, 10(8): 955–971.

Malczewski, J. 1999. *GIS and Multicriteria Decision Analysis*. Chichester: John Wiley & Sons.

Malczewski, J. 2004. 'GIS-based Land-Use Suitability Analysis: A Critical Overview'. *Progress in Planning*, 62(1): 3–65.

Malczewski, J. 2006. 'GIS-based Multicriteria Decision Analysis: A Survey of the Literature'. *International Journal of Geographical Information Science*, 20(7): 703–726.

Mallick, B., K. R. Rahaman and J. Vogt. 2011. 'Coastal Livelihood and Physical Infrastructure in Bangladesh After Cyclone Aila'. *Mitigation and Adaptation Strategies for Global Change*, 16(6): 629–648.

Mansur, A. V., E. S. Brondízio, S. Roy, S. Hetrick, N. D. Vogt and A. Newton. 2016. 'An Assessment of Urban Vulnerability in the Amazon Delta and Estuary: A Multi-criterion Index of Flood Exposure, Socio-Economic Conditions and Infrastructure'. *Sustainability Science*, 11: 1–19.

Mechler, R. and L. M. Bouwer. 2015. 'Understanding Trends and Projections of Disaster Losses and Climate Change: Is Vulnerability the Missing Link?'. *Climatic Change*, 133(1): 23–35.

Meyer, V., S. Scheuer and D. Haase. 2009. 'A Multicriteria Approach for Flood Risk Mapping Exemplified at the Mulderiver, Germany'. *Natural Hazards*, 48(1): 17–39.

Mitchell, J. K. 1989. 'Hazards Research'. In G. L. Gaile and C. J. Willmott (eds.), *Geography in America*, pp. 410–424. Columbus: Merill.

National Research Council (NRC). 2007. *Successful Response Starts With a Map: Improving Geospatial Support for Disaster Management*. Washington, DC: The National Academies Press.

Newton, A. and J. Weichselgartner. 2014. 'Hotspots of Coastal Vulnerability: A DPSIR Analysis to Find Societal Pathways and Responses'. *Estuarine, Coastal and Shelf Science*, 140: 123–133.

Paul, S. K. 2012. 'Vulnerability to Tropical Cyclone in the Southern Bangladesh: Impacts and Determinants'. *Oriental Geographer*, 53(1–2): 19–40.

Paul, S. K. 2013. 'Vulnerability Concepts and Its Application in Various Fields: A Review on Geographical Perspective'. *Journal of Life and Earth Science*, 8: 63–81.

Paul, S. K. 2015. 'Post-cyclone Livelihood Status and Strategies in Coastal Bangladesh'. *Rajshahi University Journal of Life & Earth and Agricultural Sciences*, 41: 1–20.

Paul, S. K. and J. K. Routray. 2010. 'Flood Proneness and Coping Strategies: The Experiences of Two Villages in Bangladesh'. *Disasters*, 34(2): 489–508.

Paul, S. K. and J. K. Routray. 2011. 'Household Response to Cyclone and Induced Surge in Coastal Bangladesh: Coping Strategies and Explanatory Variables'. *Natural Hazards*, 57(2): 477–499.

Pazzi, V., S. Morelli, F. Pratesi, T. Sodi, L. Valori, L. Gambacciani and N. Casagli. 2016. 'Assessing the Safety of Schools Affected by Geo-Hydrologic Hazards: The Geohazard Safety Classification (GSC)'. *International Journal of Disaster Risk Reduction*, 15: 80–93.

Pijawka, K. D. and A. E. Radwan. 1985. 'Transportation of Hazardous Materials, Risk Assessment and Hazard Management'. *Dangerous Properties of International Material Reproduction*, 5(5): 2–11.

Preston, B. L., C. Brooke, T. G. Measham, T. Smith and R. Gorddard. 2009. 'Igniting Change in Local Government: Lessons Learned From a Bushfire Vulnerability Assessment'. *Mitigation and Adaptation Strategies for Global Change*, 14: 251–283.

Preston, B. L., E. J. Yuen and R. M. Westaway. 2011. 'Putting Vulnerability to Climate Change on the Map: A Review of Approaches, Benefits, and Risks'. *Sustainability Science*, 6(2): 177–202.

Proag, V. 2014. 'The Concept of Vulnerability and Resilience'. *Procedia Economics and Finance*, 18: 369–376.

Rahman, N., M. A. Ansary and I. Islam. 2015. 'GIS Based Mapping of Vulnerability to Earthquake and Fire Hazard in Dhaka City, Bangladesh'. *International Journal of Disaster Risk Reduction*, 13: 291–300.

Smith, K. 1992. *Environmental Hazards: Assessing Risk and Reducing Disaster*. London: Routledge.

Susman, P., P. O'Keefe and B. Wisner. 1984. 'Global Disasters: A Radical Interpretation'. In K. Hewitt (ed.), *Interpretations of Calamity From the Viewpoint of Human Ecology*, pp. 264–283. Boston: Allen and Unwin.

Timmerman, P. 1981. *Vulnerability, Resilience and the Collapse of Society*. Environmental Monograph No. 1. Institute for Environmental Studies, Toronto University.

Turner, B. L., R. E. Kasperson, P. A. Matson, J. J. McCarthy, R. W. Corell, L. Christensen, N. Eckley, J. X. Kasperson, A. Luers, M. L. Martello, C. Polsky, A. Pulsipher and A. Schiller. 2003. 'A Framework for Vulnerability Analysis in Sustainability Science'. *Proceedings of the National Academy of Sciences*, 100(14): 8074–8079.

United Nations Disaster Relief Co-ordinator (UNDRO). 1982. *Natural Disasters and Vulnerability Analysis*. Geneva: Office of the United Nations Disaster Relief Co-ordinator.

Watts, M. J. and H. G. Bohle. 1993. 'The Space of Vulnerability: The Casual Structure of Hunger and Famine'. *Progress in Human Geography*, 17(1): 43–67.

Weichselgartner, J. 2001. 'Disaster Mitigation: The Concept of Vulnerability (revisited)'. *Disaster Prevention and Management*, 10(2): 85–95.

Weichselgartner, J. and J. Bertens. 2000. 'Natural Disasters: Acts of God, Nature or Society? On the Social Relation to Natural Hazards'. In M. A. Andretta (ed.), *Risk Analysis II*, pp. 3–12, Southampton: WIT Press.

White, G. F., Calef, W., Hudson, J., Mayer, H., Schaeffer, J., and Volk, D. (1958). Changes in Urban Occupance of Flood Plains in the United States (Research Paper No. 57). Chicago: University of Chicago Press, Department of Geography.

Wisner, B., P. Blaikie, T. Cannon and I. Davis (ed.). 2004. *At Risk: Natural Hazards, People's Vulnerability and Disasters*. London: Routledge.

Wolf, S., J. Hinkel, M. Hallier, A. Bisaro, D. Lincke, C. Ionescu and R. J. Klein. 2013. 'Clarifying Vulnerability Definitions and Assessments Using Formalisation'. *International Journal of Climate Change Strategies and Management*, 5(1): 54–70.

Yorke, C., F. B. Zhan, Y. Lu and R. Hagelman. 2015. 'Incorporating Evacuation Potential Into Place Vulnerability Analysis'. *Geomatics, Natural Hazards and Risk*, 6(3): 195–211.

13 Natural resource appraisal in Bangladesh

Md. Humayun Kabir

1. Introduction

Rapid population growth throughout the world, particularly after the Second World War, has brought severe stress on natural resource systems. Due to over-exploitation and increasing pressure over the natural resource systems on earth, humankind is also faced with the severe challenge of fulfilling its fundamental needs. The global population has already reached 7.35 billion in 2015 and is also projected to be over 8.5, 9.7 and 11.2 billion by 2030, 2050 and 2100, respectively (UNDESA 2015). The increasing population pressure continues to happen in the developing world, particularly in the South Asian region. Bangladesh, with a small area (148,000 km^2) accommodates a huge population (160 million) until the present (2015) and will continue to have more population in the coming future. Population density is one of the highest (over 1100/km^2) on earth, while the rapid growth of population in recent decades has brought immense pressure on natural resources, particularly land, forests and aquatic resources. In order to feed the large population size, food production over the last few decades has been enhanced manifold through the intensification of agricultural production (Rasheed 2008). Although the country's major part of the land resources is classed as a fertile floodplain, which is highly supportive of agricultural production, land resources are under tremendous stress mainly due to over-exploitation and unplanned utilization. Likewise, wetland resources are also highly stressed due to the gradual expansion of human settlements, agricultural practices and infrastructural development. Human interventions have resulted in the remarkably rapid shrinkage of wetland areas in the country. Another natural resource under severe threat of degradation are the forest resources of Bangladesh. The aforementioned reasons of human interventions are also equally responsible for the rapid reduction of forest cover. The gradual depletion of all these natural resources in Bangladesh ultimately results in significant resource scarcity and the degradation of ecological processes.

The present chapter is an attempt to unveil the urgent need for appraising the existing natural resources with increasing demand over these resources in Bangladesh and ultimately to assess their potential for development in order to meet the unavoidable needs of the country's people.

2. Need for resources appraisal

As previously discussed, the global population is gradually increasing and thereby putting tremendous pressure on scarce natural resources, particularly in the developing countries of Asia and Africa. In view of the rapid growth of the global population and the increasing demand for resources, appraisal is therefore essential, mainly with an aim to yield the greatest optimum benefits from the resource systems, to maintain essential ecological processes and to ensure the equitable share of the benefits arising from resource systems. In other words, maintenance of social equity with limited natural resources is of a priority given the current global situation.

2.1 Process of appraising natural resources

Natural resources appraisal of an area follows a series of steps, while it focuses upon resource inventory (location, distribution and areal extent of resources) (Mitchell 1979). In this regard, surveying is an important step to initiate resource appraisal process. Then, the existing resources are mapped and finally analyzed. Resource supply and demand are also important concerns in resource appraisal. While doing appraisal of resources, several questions arise with regard to resources supply that include: (a) where are the resources? (location), (b) how much exists there? (extent), (c) what is their condition? (state), (d) how will changing price affect future demand? (future scenario) (e) how will different resource uses interact to influence future supplies? and (f) What opportunities exist to improve productivity? (Mitchell 1979). In addition to the previous queries, other questions may also come in front in appraising resources (Mitchell 1979). They are: what resources should be inventoried first? How much to be spent for each type of resource? What parts of a country should be surveyed first? What inventory methods should be applied?

2.1.1 Determining supply

Supply of resources focuses upon actual existence or potential extraction, while the actual supply is concerned with actual use, referring to the geographical contribution towards the inventory of that particular resource. In case of land resources appraisal, the measuring of actual use is the inventory of the existing land use. Similar resource issues and problems arise in the case of other natural resources like forests, minerals and aquatic resources. The potential extraction of resources deals with resource capability, referring to what extent the existing resources can be improved to enhance productivity. Examples can be drawn from the inventory of land resources use in Bangladesh (Hussain et al. 2012). Although the total area until 2012 was 14.84 million hectares, the total cropped area was 14.41 million hectares. It can therefore be argued that a majority of the land resources in Bangladesh is currently being exploited for crop production. Only less than 2 percent of the total land resources are cultivable waste and current fallow.

2.1.2 Estimating demand

Demand estimation in the case of resources incorporates three phases: identification of variables, measurement of relationship and articulation of theory (Mitchell 1979). It is also argued by the researchers that demand estimation in the case of water resources is a problem particularly in identifying relevant variables (Mitchell 1979). Other difficulties faced include selection of appropriate time horizon, uncertainty and consideration of short-term variability. However, in estimating demand (in case of water resources), several other variables are easy to identify but difficult to measure. They are population estimation (number, age, sex, mortality, fertility, migration, etc.), nature of the economy (personal income, employment, industrial output, etc.), technology (e.g., existing available technology and changes influence the use of resources), societal tastes (preferences, aspirations, desires, etc.) and policy decisions (prevailing policy and changes also influence resource demand). To draw an example of estimating demand for surface water resources in Bangladesh, variables like sectors of water use, water demand for agricultural purpose, number of farming plots, farmers, etc., along with existing practices of irrigation systems, are important aspects to be assessed.

3. Natural resources appraisals in Bangladesh

Bangladesh is widely endowed with various natural resources, such as land, forests, wetlands, mineral resources, etc. Like the global trend, with rapid population growth, natural resources of different types have drastically dwindled in Bangladesh. Given the gradual rise of human population and increasing demand over these natural resources, it is pragmatic and essential to appraise natural resources for sustainable human environment. The following section presents a sequential discussion on the significance and processes of appraising different natural resources of the country.

3.1 Land resources and issues

Land is the most important and useful natural resources in any country in the world, particularly in countries where extensive floodplains associated with agro-based economic activities are predominant. At the international level, public concern on land has been more with pollution aspects of environment, for example, role of land resources as a sink (waste disposal, pollution), nature conservation (biological and genetic resources). On the other hand, at the national level (especially developing countries), the role of land resources for production is a major concern.

There also exist a significant number of issues related to land. Exploitation of land in an inappropriate way is one of them. Inappropriate land use leads to inefficient exploitation of natural resources, destruction of land resources, poverty and other social problems and even to the destruction of civilization, since land is the ultimate source of all resources and the foundation on which civilization is constructed. However, the role of society to influence land evaluation includes discouraging and forbidding the landowners to do whatever they want with their land, for various reasons. Therefore, in order to maximize the benefit from the land resources, land use planning is essential.

3.1.1 Land use

One of the most influential workers in this field is Dudley Stamp, a British geographer (1898–1966). He devised a land use classification system for Britain during the 1930s and later was a driving force in developing a system to be used for a world land use surveys. Stamp (1960) believed that a number of steps (survey, analysis and planning) were required to use land or other resources to best advantage. The first land use survey was carried out in Britain in 1930 and later in 1961 by Alice Coleman (Coleman, 1961). By 1950, the existing land use in all parts of the world was made on a uniform system of classification and notation. However, land resources appraisals started in the world through the application of aerial photography and later remotely sensed data. The major advantage of using satellites is that data collection on relatively larger areas is possible, feasible and accurate with satellite imageries. The first satellite designed for studying the earth's resources is the Earth Resources Technology Satellite (ERTS), which was launched in 1972 (Mitchell 1979). Later, the application of ERTS helped capture information on ecological patterns, geological patterns, physical relationships and land use patterns for various resource related studies, including identifying hazard and environmentally sensitive areas, monitoring pollution, crop growth and assessing wildlife. However, followed by ERTS, further versions of various latest satellite imageries have been used in assessing various natural resources particularly land resources.

3.1.2 Land capability

From the land capability map, various attributes of land as soil fertility, water availability and drainage capacity of land, etc. are known. Based on the land capability, land use is recommended to maximize the productivity and to reach the potential. The resource manager needs to know the potential or capability before making decisions. Many countries have already developed classification systems for resource capability. Sir Dudley Stamp classified agricultural lands of Britain into three categories. They include- Category I: good quality land (first class land capable of intensive cultivation, good general purpose farmland, first class land with water conditions especially favouring grass, good but heavy land); Category II: medium quality land and Category III: poor quality land.

3.2 Land resources appraisal in Bangladesh

In view of the rapid growth and high density of population, it is important to appraise the land resources of Bangladesh. The total area of the country is about 14.84 million hectares while 3.60 million hectares (around 25 percent of the total area) is not available for agriculture (Hussain et al. 2012). This significant part of the land resources is occupied by scattered homesteads, urban centres, industries, educational institutions and inhabited lands. In 1971, the net cropped area was 8.24 million hectares and reached 8.85 million hectares in 1987 with a cropping intensity of 150.73 percent (Hussain et al. 2012). The current statistics shows that the total cropped area increased to 14.41 million hectares although the net cropped area decreased to 7.94 million hectares. However, the previous discussion reveals that land resources are highly correlated with crop production in Bangladesh.

Potential for crop production (e.g., crop suitability) in the given biophysical conditions is regarded as one of the most important aspects of land resources appraisals. Agriculture is the major source of food production and prime source of raw materials for industries. Growth and development of plants are both influenced by the environmental conditions – particularly climate and soil. Therefore, the physical conditions should be adequately assessed in any regional development plan (Hussain et al.2012). The first step however would be to identify areas where climate and soil are suitable for particularly crop production. At the same time, the success or failure of crop production for the suitable areas also depends on socio-economic factors of the area that grows the crop. Other studies suggest that land resource appraisal includes the initial assessment of the biophysical potential of natural resources (Sivakumar and Valentin 1997).

Agro-ecological zoning approach presents a useful preliminary assessment of this potential. Moreover, crop zoning can be conceptualized in the context of optimizing crop production through well-judged use of land resources in order to prevent land degradation, plan for optimum use and conserve for future. The activities on land resources appraisals started in the country in 1979 under the FAO/UNDP Land use Advisory Project to make soil survey information as a basis for more rational planning of agricultural development. The physical resources database on land, soils, climate, hydrology and land suitability were used for national and sub-national agricultural research and development planning. This effort was made as a part of the five-year project entitled Utilization of Agro-Ecological Zone (AEZ) Database and Installation of GIS for Agricultural Development conducted at Bangladesh Agricultural Research Council (BARC) with the assistance of the United Nations Development Programme (UNDP) in 1996. Under this project, for the first time in the history of the country, a land resources inventory was conducted and physical attributes of soil analyzed in detail.

Geographic Information System (GIS) based maps on various characteristics of soil (soil permeability, moisture holding capacity, nutrient availability, soil salinity, consistence, drainage condition, inundation, slope, etc.) have been prepared for the entire country. The application of land resources inventory database can now widely be used by various stakeholders for overall agriculture and regional development planning. The land resources inventory (1997–2001) contains various attributes of physical characteristics of soils in Bangladesh. An

ArcView GIS-based application was developed as part of this endeavour to dynamically combine a user-specified Digital Elevation Model (DEM) with the national soil association layer to create more detailed soil association by inundation land type layer. The first step in developing this application was to refine a previously generated 300metres DEM by filling in areas of missing elevation with values taken from a 1,000 metres DEM. An avenue programme was written to assign soil association and topographically derived inundation land type designations to each 300metres grid cell. The following designations are used to specify a range of inundation depths based on the average peak water depth in Bangladesh (Table 13.1).

3.3 Wetland resources appraisal in Bangladesh

Bangladesh is largely endowed with wetland resources comprising 11 percent of the total area (16,000 km^2) of the country. Wetlands include extensive *haor* region of the northeastern part, hundreds and thousands of large *beels*, *jheels*, *baors*, rivers, lakes, ponds, ditches, etc. *Haors* are often created due to tectonic processes. The floodplain depression wetlands are called 'haors' and the deeper sections that retain water in the dry months are called '*beels*'. The *haor* consists of a few beels of various sizes. Wetlands of Bangladesh can be divided into three categories based on the hydrological and ecological considerations. They are salt water or saline or brackish, freshwater and man-made/artificial (e.g., ponds). However, the following section presents

Table 13.1 Distribution of land type by inundation in Bangladesh

Land type	Characteristics	Areas (hectares)
Highland (H)	Land above normal inundation	4,199,952
Medium highland (MH)	Land normally inundated up to 90 cm deep This type is further divided into: MH-1: Inundated up to 30 cm deep MH-2: Inundated 30–90 cm deep	5,039,724
Medium lowland (ML)	Land normally inundated up to 90–180 cm deep	1,771,102
Lowland (L)	Land normally inundated up to 180–300 cm deep	1,101,560
Very lowland (VL)	Land normally inundated deeper than 300 cm	193,243

Source: Adapted from UNDP and FAO 1988

an overview of the existing state of wetland resources in Bangladesh, issues concerning wetland degradation, significance of wetland conservation and the case of a Ramsar Site (Tanguar *Haor*).

Globally, wetland resources are given increasing attention particularly since the first congregation in the city of Ramsar, Iran in 1971. The first effort on earth to conserve the wetlands of international importance was made through the establishment of the Ramsar Convention Bureau. It defined the term 'wetland' as

> the areas of marsh, fen, peat land or water, whether natural or artificial, permanent or temporary, with water that is static or flowing, fresh, brackish or salt including areas of marine water, the depth of which at low tide does not exceed six metres can be defined as wetlands (Ramsar Convention Bureau 1971).

The Bureau for the first time brought the ideas on the application of wise-use principle of wetland resources. Until now, around 2,000 sites of the world are declared as Ramsar sites. In Bangladesh, the first wetland declared as a Ramsar site is the Sunderbans in 1999. Later in 2003, Tanguar Haor was declared as the country's second Ramsar Site (Kabir and Amin 2007).

Wetlands play very important roles in different sectors of Bangladesh including for ecosystem, economy, agriculture, fisheries, forestry, etc. They offer habitats to a large variety of wildlife and safe nesting to waterfowls. Wetlands also offer a favourable breeding environment to large number of resident and migratory birds. They are also contributing as a huge source of fishery resources that maintain protein intake for humans. Wetlands support a very favourable condition for cattle grazing and they enrich soil fertility. However, the existence of the wetlands all over the country is rapidly dwindling. Increasing population pressure and the expansion of human settlements, widespread poverty, commercial and over-harvesting of fishery stock (in the past by the leaseholders), inefficient exploitation of wetland resources by the users, expansion of agricultural activities, overuse of wetland forests as fuel, loss of habitats and degradation, pollution due to agro-chemicals, conflicts between the locals and leaseholders and lack of awareness among the locals and lack of locals' participation in conservation practices are the various reasons for wetland loss and degradation in the country. Considering the existing state of the ecological processes and their importance, some of the wetlands, particularly in and around Dhaka (the river Buriganga, Turag, Shitalakkya, Balu, Gulshan-Baridhara Lake), have been declared as Ecologically Critical Areas (ECAs).

Given the state of over-exploitation and population pressure, wetland resources are under severe stress. In order to support a large population size living in and around the resource systems, the appraisal of wetland resources is essential. The appraisal of wetland resources includes evaluation of the potential supply of natural resources from the existing wetlands and assessing the demand over the aquatic natural resources of Bangladesh. In the first step, the entire wetlands areas have to be identified and mapped. The biophysical and ecological characteristics of the existing wetlands have to be critically analyzed. As mentioned earlier, due to the country's diversity of landscapes, saline/brackish (mangrove wetlands/sunderbans) and freshwater (Hakaluki Haor, Tanguar Haor, etc.), wetlands exist in different locations. Due the diversity of the nature of these wetlands, the biophysical and ecological characteristics are also substantially different. Millions of inhabitants live in and around these wetland resource systems. The demand of the local people living in and around the wetlands has to be assessed with due importance. In most cases, locals' livelihoods are entirely dependent on wetland resources. Thus, the management authority and local people are often in conflict, while the locals are often blamed for resource depletion and illegal harvesting. The local people being very poor have to depend on the wetland resources (for fishing, collection of fuel wood, trapping of fish spawns, etc.). On the other hand, corrupt and inefficient management of authority, involved in the blame-game on the locals, is also reported as an agent of resource degradation. For example, Tanguar *Haor*, the country's second Ramsar site, has been managed by the district administration, completely excluding nearly 100,000 local people from resource use. Local people being entirely dependent on haor resources often go for illegal harvesting and face punishments of different scales. Meanwhile, the Tanguar Haor Master Plan has been prepared and, until now, the locals have not been incorporated in the conservation practices. Such situations also exist in various wetland resource systems (for example, Hakaluki *Haor*, Arial *Beel*, etc.).

Therefore, wetland resources appraisals require incorporating all these issues and challenges. In order to develop and conserve wetland resources in Bangladesh, activities required include stopping the development activities of the identified sensitive wetlands, integrating mitigation measures of the environmental parameters in all existing development projects on wetlands, restoration of the essential ecological processes in the degraded wetlands and, finally, integrating the local people in the conservation practices so that they are not detrimental to the wetland resource systems.

3.4 Forest resources appraisal in Bangladesh

Like wetland and water resources, Bangladesh has also been endowed with extensive forest resources of various types and extents. Over time, mainly due to uncontrolled and illegal exploitation and lack of proper management, the forest resources have rapidly dwindled since the country's independence. The population pressure around the forest resource systems is always seen as a major issue of such degradation. The forest types of Bangladesh include hill forest, freshwater swamp forest (e.g., Ratargul swamp forest, Sylhet), *sal* forest (e.g., Madhupur), mangrove forests and coastal plantation forests (Table 13.2).

It is reported that although the forest lands remain around 17 percent, the forest/vegetation cover has reduced to 6 percent (Hasan 2016). Given the country's rapid growth of population and increasing demand over forestry resources (e.g., timber), forest resources are under severe stress. Along with human occupation and illegal logging in all the forests, poor and corrupt management practices are also responsible for such a situation. From an ecological perspective, it is essential to ensure the existence and continuation of 25 percent of the total areas as forests. However, forest resources appraisals are essentially important in the sense that under proper management, these resources have to be assessed for the objectives of conservation and satisfying the needs for present and future. Like wetland resources, the first step to appraise forest resources includes identification and mapping of the geographical location and extent of various forest types. In this case, the entire forest resources need to be mapped with the latest technologies (e.g., GIS/Remote Sensing (RS)). High resolution satellite data would be very much effective in detecting the existing state of forest cover, their location, extent, types and height of vegetation, abundance, etc. Existence of local communities (number, density, livelihood

Table 13.2 State of major forest types in Bangladesh

Forest type	Total area (hectares)	Area managed by Forest Department (hectares)
Hill	1,377,000	670,000
Sal	120,000	120,000
Mangrove	601,700	601,700
Swamp	23,000	23,000
Coastal plantation	200,000	200,000
Total	2,321,700	1,614,700

Source: Prepared by Author from Paul (2016) and DoF (2015)

patterns, etc.) and their demand (e.g., non-timber forest products) over these resources have to be critically assessed. The future demand can also be calculated with existing demand and growth. Thus, the potential sites have to be identified for conservation and development.

3.5 Energy resources appraisal in Bangladesh

Adequate supply of energy and electrical power is regarded as an indicator of development. But Bangladesh is recognized as a country of energy and power crisis. At least 60 percent of energy is still obtained from biomass energy, especially in the rural areas (Kabir and Endlicher 2012). Other supply mostly takes place through non-renewable or conventional energy sources. Indigenous gas (available within the country), oil (mostly imported petroleum) and coal (currently being mined within the country and also imported) are the major sources of primary commercial energy in Bangladesh. The country's power is generated mostly with conventional fuel (82 percent indigenous natural gas, 9 percent imported oil, 5 percent coal) and renewable sources (4 percent hydropower and solar). At the moment, it is claimed that around 70 percent people are connected with the public utility grid (BPDB 2015). Until the present (2015), the country has achieved an installed capacity of around 10,000 MW. But more than 40 million people are still living in the off-grid areas. The unreliable electricity has been directly linked to loss of productivity and hence to the lowering of the economic growth rate. In view of the rapid decline of natural gas (a major ingredient to produce electricity) and the government's vision to provide electricity to all by 2020, the country is going to encounter a massive challenge to overcome the situation. The following section presents an overview of the existing state of non-renewable energy resources and potential renewable energy sources given the state of power generation and supply in Bangladesh. It also highlights the significance of appraising energy resources, particularly the potential application of solar photovoltaic systems given the country's power crisis situation and future demand.

3.5.1 Non-renewable energy resources

The major sources of non-renewable energy of the country are natural gas and coal. The entire reserves of exploitable indigenous fossil fuels, with the exception of the coal reserve, are located in the eastern part of the country. This results in a gap of commercial energy supply between the east and the west (Hussain and Badr 2005). The trend is mainly due the increased consumption of indigenous natural gas and imported oil. Among all other energy sources, natural gas contributes a significant part (nearly 70 percent) to the total primary energy and plays a vital role in the growth of the national economy. The dependence on natural gas is due to the fact that Bangladesh does not have any other energy resource in significant quantities. It was earlier projected that the contribution of natural gas in electricity generation drops from 80 percent to 61 percent by the year 2015. Gas based electricity generation will drastically be reduced (less than 1 percent) by the year 2030 (Kabir et al. 2010). Power plants, fertilizer factories, other industries (e.g., brick kilns, tea processing plants, steel mills and textile factories), commercial organizations (e.g., offices and business centres) and the domestic sector are the end users of natural gas in the country. The gas consumption in the year 2004–05 was 487 bcf, which on a daily basis is 1,334 MMCFD (million cubic feet per day). Gas consumption has been growing at an average rate of 7 percent per year for the last two decades (CES 2006). Ahmed (2014) reported the highest daily gas production in May 2014, at 2,350.4 MMCFD. In Bangladesh, as many as 22 natural gas fields have been discovered up to June 2004 (GOB 2005).

The major coal deposits of the country are located at Jamalganj in Jaipurhat District, Baropukuria and Phulbari in Dinajpur District and Khalashpir in Rangpur District. Among them, the coal reserve in the field of Jamalganj has been estimated to be nearly half of the total reserves in the country (Hossain 2009). However, there are five coal fields in Bangladesh with nearly 2.55 billion tons of reserves (Energy and Power 2009). The Barapukuria coal mine was discovered in 1985 by the Bangladesh Geological Survey (GSB), and the mining activity is taking place only at this location. The Phulbari coal deposit of Dinajpur District was discovered during the second half of the 1990s by the BHP, an Australian mining company. Later on, the Asia Energy Corporation, a UK based Global Coal Resources Plc., was awarded the licensing agreement to operate the reserve. Recently, there was an unprecedented event with the attempt of coal extraction at Phulbari Upazila. The project, being an open-pit mine, if operated would create serious environmental impacts on the local area and could dislocate around 50,000 people from the locality. Bangladesh has not been able to make remarkable advancements in exploiting oil from the only reserve discovered in December 1986, located at Haripur in the District of Sylhet. There is an estimated reserve of 1.4 mtonne of which 0.84 mtonne are believed to be recoverable (Hussain and Badr 2005). Due to the reduction of pressure in the field and the influx of water into the oil zone, exploitation of oil has not taken place. In 2006, 2 percent of diesel and 6 percent of furnace oil were used for electricity generation.

3.5.2 Renewable energy resources in Bangladesh

As mentioned earlier, despite the availability and enormous potential of renewable energy, there has not been any significant progress in the promotion and development of renewable energy technologies, mainly due to highly expensive installation devices, high maintenance costs and lack of strong political commitment. The country, being largely dependent on agriculture, produces a huge amount of biomass, which is the only source of energy supply to the remote rural villagers. Another renewable energy technology used in electricity generation is hydropower commissioned in 1960s in the south-eastern hilly region known as Kaptai Dam. Until present, a few assessment projects on solar and wind energy have been conducted, and a large number of standalone Photovoltaic (PV) systems have been installed in the rural areas.

Bangladesh, though a small country, has numerous potential sources of renewable energy, for instance biomass, solar energy, hydropower, wind and tidal energy. Energy supply in the country is still made mostly from the renewable sources in the sense that a large part of the inhabitants in the rural areas receives energy from traditional biomass. The sources are mainly cow dung, domestic wastes, jute stick, rice straws, twigs, etc. The second largest renewable energy deployment is hydropower, which generates around 5 percent of the total consumption. Among the new renewables, which are reported to have greater potential, are solar and wind energy (Table 13.3). But a very insignificant number of studies have been undertaken to explore the potential of tidal, wave and ocean thermal energy resources. A number

Table 13.3 Renewable energy prospects in Bangladesh

RES type	Capacity (up to December 2016)	Theoretical potential
Wind	1 MW	2,000 MW
Hydro	230 MW	672 MW
Solar PV	200 MW approx.	50,436 MW
Solar thermal	3,000 m^3	20×10^6 m^2
Biogas	1 million m^3	$3,675 \times 10^6$ m^3

Source: Based on Alam et al. (2003), IDCOL (2017)

of academic (autonomous), government, non-government and private institutions have been working on the development and promotion of renewable energy technologies in Bangladesh. But due to a number of inevitable reasons (e.g., lack of strong political will, high initial investment, high cost of devices, maintenance, good quality assessment data and so on), outstanding progress has not been achieved in harnessing these renewable energy sources.

Hydropower: In the global context, the installed capacity of hydropower has just reached 1,000 GW (REN21 2015). Bangladesh, being almost a flat country does not possess extensive potential in hydropower resource except some small hydro-projects. The Karnafuli Multipurpose Hydroelectric Project (KMHEP) is the first renewable energy development project in the country with an installed capacity of 230 MW. Since its inception, the plant has been operated by the Bangladesh Power Development Board (BPDB) and contributed approximately 5 percent of the total national electricity generation in 2003 (Hussain and Badr 2005). BPDB is considering a 100 MW capacity extension of this hydropower plant. The additional energy will be generated during the rainy season when most of the year's water is spilled. Apart from Kaptai dam project, the BPDB has identified two other prospective sites for hydropower generation. They are at Sangu (100 MW) and Matamuhuri (75 MW) Rivers. Small and micro hydroelectric potential in the country is around 200 MW, and thus the aggregate generation could be 745 MW (SWERA/RERC 2007).

Wind Energy: Wind energy utilization in Bangladesh is still in the early stage of application except a few pilot projects and wind-battery hybrid plant in the coastal parts of the country. The generation of wind-generated electricity started practically from zero in the early 1980s to more than 10 TWh per year in 1996 (Islam et al. 2006). The Bangladesh Meteorological Department (BMD) has wind speed measuring stations in towns and cities at heights between 5 and 10 metres above ground level. Later on, several studies have been conducted by the public departments, national level NGOs and international agencies. Several locations have already been assessed to evaluate the wind energy potential of the coastal region. Variation of wind resource has been observed among the different locations in the region. In addition, seasonal variation of wind resource prevails in the country, with a potential during the months of April-September, and a very weak potential during rest of the year.

Biomass Energy: Remote rural areas of different countries worldwide still depend on biomass energy. Sarkar et al. (2003) reported that the contribution of biomass energy as source of energy to rural

Bangladesh is still significant (~60 percent). In this sense, renewable energy contribution is much higher compared to other forms of energy. The rural people, being very poor, depend on woods, residues of agricultural products, animal manure, etc. Another way of generating power from biomass is gasification of woods, straws, crop residues, etc., which is becoming very popular worldwide. In Bangladesh, very few biomass gasification plants have been installed.

3.5.3 Potential application of solar energy

As previously mentioned, the non-renewable energy sources are rapidly declining and there is immense scope to exploit country's abundance of solar energy (Kabir et al. 2010; Kabir and Endlicher 2012). The Renewable Energy Policy (REP) in the country has just been approved, which can be seen as a major initiative to promote the application of solar PV systems. At this moment, several national level NGOs, particularly Grameen Shakti, are receiving grants from the government owned company Infrastructure Development Company Ltd. (IDCOL), which has been disseminating solar home systems in the off-grid rural areas. Until present (2016), a total of six million home systems have been installed with a capacity of nearly 200 MW of electrical power (IDCOL 2017).

In order to make an appraisal of solar PV based electricity generation in the megacity of Dhaka, Kabir et al. (2010) made a potential assessment of the rooftop application of solar PVs. Through an Object Based Image Analysis (OBIA) of the Quickbird satellite image of Dhaka 2006 (with 0.6 metre resolution), well-illuminated roof areas of Dhaka were identified and calculated (Table 13.4). The Quickbird

Table 13.4 Well-illuminated roof areas from Quickbird Image 2006 of Dhaka City

Various objects classified	Area calculated (km^2)	Percent of the total area calculated
Bright rooftops	10.554	7.86
Informal settlements	9.646	7.18
Vegetation	53.364	39.74
Water bodies	9.583	7.14
Others (roads, open lands, shadow, etc.)	51.135	28.98
Total	134.282	100.0

Source: Kabir et al., 2010

Table 13.5 Solar PV based power generation potential in Dhaka

Bright rooftop (km^2)	Bright rooftop (m^2)	Actual roof area for PV installation (m^2)	GHI of Dhaka ($kWh/m^2/day$)	Each Module (75 W) output (W)	Area needed/ module (m^2)	No. of modules to be installed	Potential electricity generation (MW)
10.554	10.554 × 10^6	5.277 × 10^6	4.20	75 × 4.20 × 0.6 = 189	1m × 0.8m = 0.8	6,596,250	1,246.69*

Source: *The potential generation can be much higher based on the high module capacity (e.g., 210 Wp). Based on Kabir et al. (2010)

scene of Dhaka provides immense heterogeneity with buildings of various characteristics, sporadic roads, vegetation, water bodies, open spaces, etc. It is evident that all rooftops are not appropriate for PV applications (e.g. rooftops receiving less sunlight due to building shades and vegetation cover). Although the area of the Quickbird scene is 431 km^2, the area of the wards (134.282 km^2) of the former Dhaka City Corporation (DCC) has been considered for rooftop identification and calculation. In order to avoid much complexity and error, the areas with the most building density (DCC wards area) have been separated from the Quickbird image based mainly on field experience and visual observation. The other areas of the Quickbird image mostly cover water bodies, agricultural lands, vegetation, roads and buildings with less density. It can be considered that in a very heterogeneous city like Dhaka, the calculation of bright rooftops for PV application within the DCC ward areas would substantially contribute as very useful input data. However, the calculation of bright rooftops within the DCC wards area shows that the megacity of Dhaka offers 10.554 km^2 bright roof areas. This figure can be considered as a conservative estimate. The application of solar PV systems on these bright rooftops can generate more than 1,000 MW of electricity (at 10 percent efficiency with 75 Wp modules), preferably through gird connected PV systems. The potential electricity generation can be substantially high (>1,500 MW) with the installation of solar modules with high capacity (e.g., 210 Wp) and efficiency, which would sufficiently meet Dhaka City's existing power demand. The study revealed that at least 50 percent of Dhaka's (1,000 MW) electricity demand can be supplemented through the application of solar PVs on the well-illuminated rooftops (Kabir et al. 2010; Kabir and Endlicher 2012; Kabir and Endlicher 2013).

4. Discussion

In view of the increasing population pressure on scarce natural resources in Bangladesh, there is an immense need to appraise the existing natural resources with an aim to ensure the continuous supply of life-support services from these resource systems. The previous sections have highlighted the state of different components of natural resources in the country and presented an argument on how the existing natural resource systems can be expanded through proper appraisals. However, due to the diversity of natural resources, the approach and methodology of appraisals are likely to be substantially different. While approaching such endeavours, different aspects of the natural resource systems (land, water, wetland, forest and energy) in Bangladesh have to be critically examined. Important variables to assess concerning natural resources may include various biophysical (extent, location, physical condition, etc.), and socio-economic aspects (demand and supply, number and distribution of population, pattern of livelihoods, economic activities, etc.). Inventorying of the existing resource use is the first phase to approach resource appraisals, which would ultimately be important input to the resource managers. Geographical elements, such as scales (local, regional and national) of various resource inventories, identification of the spatial distribution/characteristics based on field observation or preparation of maps showing different resource systems through GIS/RS technologies would substantially contribute in the process of resource appraisals. Therefore, knowledge concerning geographical aspects (spatial parameters such as interaction between communities, between people and resource systems, spatial organization, other biophysical characteristics, etc.) area key to successful appraisals of natural resources in Bangladesh.

5. Concluding remarks

With the rapid rise of global population, the human demand for natural resources (land, forests, wetlands, energy, etc.) is outpacing supply. Bangladesh, a populous country, currently faces enormous challenges of maintaining adequate natural resources (land, water, forests, minerals, etc.). In the case of land resources, the policy agenda of the government should be to ensure rational land use using different types of land in ways best suited to their potential, improvement of land management in agriculture and forestry so as to secure higher productivity, avoidance of land degradation, soil erosion, forest clearance, pasture degradation, etc. so as to conserve resources for the future. In order to guide decisions on the previous discussion, reliable data is essentially needed. At the same time, scientific research is to be carried out to advance knowledge on which improvements in assessment can be made. Meanwhile, such efforts have been initiated, although they do not seem enough. In order to ensure sustainable management of the country's limited natural resources, an effective appraisal in a scientific manner is essential.

References

Ahmed, R. 2014. 'Use of Natural Gas in Bangladesh'. *The Daily Observer*, 16 November.
Alam, M., A. Rahman and M. Eusuf. 2003. 'Diffusion Potential of Renewable Energy Technology for Sustainable Development: Bangladeshi Experience'. *Energy for Sustainable Development*, VII(2): 88–96.
BPDB. 2015. *Annual Report of the Bangladesh Power Development Board*. Dhaka: Government of the People's Republic of Bangladesh.
CES.2006. *Productivity Improvement in Industry Through Energy Efficient Programme*. Working Paper Submitted to the GTZ Project on Sustainable Energy.
Coleman, A. L. 1961. 'The Second Land-use Survey, Progress and Prospect'. *Geographical Journal*, 127: 168–186.
DoF. 2015. *Annual Forestry Report, Department of Forests*. Dhaka: Government of the People's Republic of Bangladesh, Ministry of Environment and Forests.
Energy and Power.2009. 'Fiver Coal Fired Power Plants'. *Planned, Energy and Power, a Fortnightly Magazine*, 6(I24), July: Dhaka.
GOB.2005. *Bangladesh Economic Review 2005*. Finance Division, Ministry of Finance, and Government of the People's Republic of Bangladesh.
Hasan, S. R. 2016. *Wildlife and Forests Acts in Bangladesh*. Paper Presented in the Seminar on World Environment Day Jointly Organized by the Department of Environment and the Department of Geography and Environment, University of Dhaka, 3 June.
Hossain, M. M. 2009. 'Coal Gasification in Jamalganj'. *Energy and Power*, A Fortnightly Magazine, 6(23), May: 9–14.
Hussain, A. K. and O. Badr. 2005. 'Prospects of Renewable Energy Utilization for Electricity Generation in Bangladesh'. *Renewable and Sustainable Energy Reviews*, 11(8): 1617–1649.
Hussain, S. G., M. K. A. Chowdhury and M. A. H. Chowdhury. 2012. *Land Suitability Assessment and Crop Zoning of Bangladesh*. Dhaka: Bangladesh Agricultural Research Council.
IDCOL. 2017. *Annual Report*. Dhaka: Infrastructure Develop Company Limited, Government of the People's Republic of Bangladesh.
Islam, M. R., M. R. Islam and M. R. A. Beg. 2006. 'Renewable Energy Resources and Technologies Practice in Bangladesh'. *Renewable and Sustainable Energy Reviews*, 12(2): 99–343.

Kabir, M. H. and S. M. N. Amin. 2007. *Tanguar Haor: A Diversified Freshwater Wetland*. Dhaka: Academic Press and Publishers Library.

Kabir, M. H. and W. Endlicher. 2012. *Exploitation of Renewable Energy in Bangladesh: Power Supply and Climate Protection Perspectives*. Dhaka: A H Development Publishing House.

Kabir, M. H. and W. Endlicher.2013. 'Supplementing Electricity Through Solar PV Systems'. In A. Dewan and R. Corner (eds.), *Geospatial Perspectives on Urbanization, Environment and Health*, pp. 203–218. Dordrecht, Heidelberg, New York and London: Springer Science & Business Media.

Kabir, M. H., W. Endlicher and J. Jaegermeyr.2010. 'Calculation of Bright Roof-tops for Solar PV Applications in Dhaka Megacity, Bangladesh'. *Renewable Energy*, 35(8): 1760–1764.

Mitchell, B. 1979. *Geography and Resource Analysis*. London: Longman Group Limited.

Paul, A. R. 2016. *Wildlife Conservation and Management of Bangladesh*. Paper Presented in the Seminar on World Environment Day Jointly Organized by the Department of Environment and the Department of Geography and Environment, University of Dhaka, 3 June.

Rasheed, K. B. S. 2008. *Bangladesh: Resource and Environmental Profile*. Dhaka: AH Development Publishing House.

REN 21. 2015. *Renewable Global Status Report: 2015 Update*. Paris: Renewable Energy Network for the 21st Century (REN21 Secretariat).

Sarkar, M. A. R., M. Ehsan and M. A. Islam. 2003. 'Issues Relating to Energy Conservation and Renewable Energy in Bangladesh'. *Energy for Sustainable Development*, 7(2): 77–87.

Sivakumar, M. V. K. and C. Valentin.1997. 'Agro Ecological Zones and the Assessment of Crop Production Potential'. *Philosophical Transactions of the Royal Society of London B: Biological Sciences*, 352(1356): 907–916.

Stamp, L. D. 1960. *Applied Geography*. Hardmondsworth: Penguin.

SWERA/RERC. 2007. *Final Report on Solar and Wind Energy Resource Assessment (SWERA) – Bangladesh*. Supported by United Nations Environment Programme (UNEP) and Global Environmental Facility (GEF), Prepared by Renewable Energy Research Centre, University of Dhaka, Dhaka.

UNDESA. 2015. *World Population Prospects*. New York: United Nations Department of Economic and Social Affairs.

UNDP and FAO.1988. *Land Resources Appraisal of Bangladesh for Agricultural Development, Technical Report-2, Agroecological Regions of Bangladesh*. Rome: United Development Programme and Food and Agriculture Organization of the United Nations.

14 Literature reviews
Searching for the deep roots of geographical knowledge

Nandini Sanyal

1. Introduction

The chapter discusses the importance of literature reviews in geographical research and discusses briefly the different techniques used in conducting one. The narratives and arguments given in the chapter indicate that research design, development of data collection instruments and data analysis plans are primarily dependent on an overall understanding of the subject matter in question. It also indicates that students and professionals undertaking geographical research need to know the wide breadth of knowledge of the subject matter so that they can efficiently fit their own piece of work within a broader knowledge framework. They need to know the theoretical frameworks that exist within geography (e.g., attributes like scale, space and spatial considerations in defining patterns, relationships and networks) and also frameworks that are borrowed by geographers (e.g., from post-modernism, post-structuralism, political ecology, feminism, sustainable development concepts, etc.) to produce good academic works. The chapter also argues that a lack of background understanding about the area of interest or limited background knowledge may conceptually confine the researchers in outlining the epistemological boundary of their research, resulting in their inability to develop an ontological structure, including identifying methodological procedures.

2. What is a literature review?

A literature review is a description, concise summary and critical analysis of the relevant reference research works in the field related to a particular research topic. In brief, it is a process of surveying scholarly literature, such as books, scholarly articles and any other sources, analyzing and documenting of what is already known about a particular topic, area of research or theory. Literature review is conducted for any purpose, either as a part of an academic dissertation or for constituting the background part of a research article or in the context of a standalone research review. It informs the current state of knowledge on a topic accumulating/emerging from the earlier research works done by other authors (Green et al. 2006). Essentially, it is not any empirical work/primary research work, rather it is a description and interpretation of the findings derived from other's works. The literature review informs the broader background of the topic within the wider subject field and the author sets his own particular research ground/position/context in the light of/in the midst of them. As a whole, it provides an impression of a range of aspects, like what has been said in the literature, i.e., the main issues it has focused on, who are the key writers, the central theories and hypotheses and the questions being asked about suitable methodologies.

The author establishes a connection between the referred scholarly literature and his own research position, locating his own work in relation to the perspective/the context of other researcher's works. The author here shows how the works of other authors are relevant to his own research context and how their works have influenced and supported his own position (Ridley 2012). It demonstrates how the present study fits in the larger field of study. The process of literature review involves a number of activities, such as searching and identifying relevant scholarly literature on a topic, organizing, reading, interpreting and analyzing them to derive up to date information with a view to answering the research questions and finally synthesizing the scholarships as a written document (Garrard 2014; Onwuegbuzie and Rebecca 2016). Hart (1998) rendered a more functional definition of the literature review, which focuses on consulting with the existing literature both published and unpublished to focus on different specific aspects/nature/characteristics of the topic and reveal how its investigation should be performed. Following his definition, the prime task of literature review is to illuminate the topic with a view to unearth its particular/certain aspects by answering pertinent research questions.

In this sense, a literature review can be conducted as a method of research, which introduces a number of techniques for analyzing, interpreting and synthesizing other people's research work; this forms the basis for another goal, such as the justification for future research in the area. It introduces and provides examples of a range of techniques that can be used to analyze ideas, find relationships between different ideas and understand the nature and use of argument in research. Knopf (2006) defines a literature review as a combination of two distinct elements, firstly, it contains the summary of the findings evolved from the previous research efforts and, secondly, it should incorporate the author's own opinion about the findings and other attributes. The author needs to add some value to it by making some conclusive judgments, like how the study was, about the overall quality, how the methodology was done, what was its limitations and, furthermore, the authors are to express their view, i.e., how reviewed literature is relevant to the author's present study and how it supports making any judgment or argument when analyzing the present topic. These special attributes of literature review makes it different from literature surveying, such as the annotated bibliography (Emerald Group Publishing).

3. Usefulness of doing literature review

Through literature review, the researcher broadly searches out and consults with what has been previously done on a particular topic and its associated subject area; indeed the key purpose of any literature review irrespective of academic disciplinary divisions should entail retrieving information and thus acquiring knowledge about any topic, thereby creating ground for the advancement of it (Onwuegbuzie and Rebecca 2016). Literature review, conducted for any purpose, has as its most important aim to provide an account of the state of the art on that particular topic and its wider subject field. It identifies the gaps in the knowledge where the proposed study could contribute, and it has the prospect of paving a foundation for new research. Thus, the process plays an important role in stimulating further advancement of knowledge production.

A literature review enables a researcher to be informed about both the methods and history (Knopf 2006). It places the researcher in the historical context of and the recent developments of the wider subject area and it enables a researcher to be acquainted with a particular field of study, including theories, vocabulary, key variables and phenomena, and enables them to look at all the corners of a research design, starting from topic selection to writing stage; it also contributes in increasing the personal skill of a researcher and increasing the awareness of the students (McCabe 2005). The following table (Table 14.1) summarizes a range of reasons why the authors conduct literature review. The common causes are categorized based on their specific focus and relevance.

Table 14.1 Necessity and purpose of conducting a literature review

A literature review for selection of research topic and to determine research focus
- Select, reselect and refine the topic and justify the significance of the proposed topic. Gaining new, clear understanding on the topic of interest
- Prior understanding of the actual research and assess the 'research ability of the topic'
- Help to identify major aim and objectives of the present research and formulate appropriate research questions for answering
- Concentrate on particular areas and enable progressive narrowing of the topic
- Define and limit the scope of the research
- Identify key researches on a topic and get information about their authors and sources
- Advance new ideas, thoughts and perspectives out of present research and make new contributions to the intended study
- Identify reasons to avoid duplication of research work.

Research epistemology/theory and methodology related reasons
- Find out the existing methods used by different authors and to be acquainted with various research designs, data collection and data analysis techniques
- Identify the theoretical, conceptual and other kinds of research frameworks
- Help to include discussions of relevant theories and concepts which set the theoretical foundation for the proposed topic. Also help to formulate the overarching theory
- In some ambitious instances, a literature review augments theory development
- Provide a context for describing, elaborating and evaluating the new theory

Acquiring knowledge about wider subject area and enhancing personal skills
- Make the researcher acquainted with body of knowledge covering the wider field
- Provide an overview on both the historical and current context of the present study. Help to inform about the contemporary concepts/approaches, ideas, terminology, vocabulary, jargon and other various phenomenon relating to the topic
- Identify gaps in the literature relating to the problem. Suggest ways of filling up the research gaps
- Identify gaps in previous research activities and limitations; also to identify further research need
- Provide historical background of the proposed research
- Receive indications to introduce innovative ideas and thoughts to the present research
- Help to enhance intellectual capability and practical skills, comprehensive thinking about the topic and its scope, methodology, analysis and interpretation techniques and overall structuring of the research
- Means of measuring research knowledge of a researcher; increase students' awareness of research through the review process

Table 14.1 (Continued)

Contextualizing the present research
- Contextualize the present research within the wider knowledge domain
- Assist to provide information for base arguments
- Provide latest source of information

Application of literature review
- Discover new knowledge for decision-making, developing standard and guidelines
- Generate new ideas and thoughts that can be used for the present study
- Describe empirical evidence for theoretical hypothesis
- A bibliography or list of sources are derived through a literature review

Source: Levy and Ellis (2006), Baumeister and Leary (1997), Hart (1998), Webster and Watson (2002), Leedy and Ormrod (2005), Levy and Ellis (2006), Burns and Grove (1999), Coughlan et al.(2007), Carnwell (1997), Knopf (2006), McCabe (2005)

4. Types of literature review

4.1 Traditional or narrative literature review

A narrative or traditional review is the most common type of literature review practiced in social science disciplines. It provides a broad overview of what is already known about a topic and the subject area. It involves a process of summarizing the findings of scholarly materials and finally synthesizing the scholarships together to draw a conclusion about the topic. In the traditional review pattern, most of the surveyed materials are needed to go through some form of critical evaluation and analysis. In comparison to other kinds of literature reviews, such a systematic review or meta-analysis of the common characteristics of a narrative review differs, many of them are not even explicit or well defined in their application, for example, it does not mention clearly the inclusion and exclusion criteria of the individual empirical or other kind of research into the present study, the literature searching process, number of studies included and the methodological rigor (Cronin 2008; Jesson 2011). Considering the main objective and focus in the review process, Onwuegbuzie and Rebecca (2016) categorize narrative in four divisions, such as (a) general literature review, (b) theoretical literature review, (c) methodological literature review and (d) historical review. For example, the methodological review provides descriptions of different methods and research designs used in different studies. A narrative literature review is invaluable when one is attempting to link together many studies on different topics, with a view to interpret in a new fashion either for the purpose of supporting a stand or making interconnections among them (Baumeister and Leary 1997).

4.2 Systematic literature review

In contrast to the narrative or traditional review, the systematic review has been distinguished on the grounds that it is more rigorous and more transparent regarding its conceptual approach, methodological consideration, qualitative approaches and more free from different kinds of biases (Petticrew 2001). The systematic review is generally conducted to answer very specific questions. It also deliberately identifies documents maintaining inclusion/exclusion criteria, critically evaluates them to maintain strict rules and procedures, and summarizes scientifically all of the selected research about a clearly defined research problem. It is widely popular among researchers and policy makers, academics and professionals of biomedical sciences, health care providers and professionals of other scientific fields who must be aware of the most recent development of their respective fields.

4.3 Meta-analysis

Meta-analysis is a kind of systematic review, which is mainly a statistical analysis technique. It is a process of combining huge amounts of quantitative data and conducting a statistical analysis for integrating the findings.

5. Sources of literature

The sources of literature for conducting a review is not limited to a particular kind of information source, rather it can extend from textbooks to current web-based resources.

Textbooks published by well-known publishers are a good place to start while searching out relevant literature on a topic and the wider subject area. Generally, university libraries offer the facilities of accessing hard copy books. The libraries of third world countries face a large difference in the number of textbooks compared to the developed countries. Nowadays, in developed countries, many textbooks are available to read through e-book readers. Dictionaries, encyclopedias, thesis and directories present the basic information and definition of a particular phenomenon.

Research articles published in reputed peer reviewed journal are a most dependable source to consult with as these articles are to go under scholarly scrutiny through the peer review process by the guidance of other knowledgeable and experienced reviewers (Webster and Watson 2002). The peer review process maintains high standards before publishing each article. Recently published journal articles are the most current source of information on a particular topic and it keeps changing periodically, on the other hand, earlier published articles help to know the previous state of knowledge on the topic. In addition to this, the references list at the end of a journal article provide the opportunity to a writer to look into each individually for a new interpretation if needed. Nowadays, many universities subscribe to both hard copy and electronic format journals. Non-peer reviewed articles could also be searched for source information.

In addition, other sources like encyclopedias, dictionaries, guides and handbooks are helpful as supporting sources. Conference proceedings, academic theses, working papers of different researchers also could be a good source of what might not be published yet but that also indicate their current initiatives of research. Recently, the internet has been the most popular sources of diversified kinds of information. Search engines or information gateway facilitates accessing information about a particular topic. Search engines are large databases of websites Younger (2004) connected by software called spiders or web crawlers. When keywords are typed into a search engine like Google, the search engine matches these keywords with its directories of information. Considering other relevance and other criteria, the search engine shows the best matching websites as the search result. Google Scholar is another popular search engine used for scholarly internet sources. Google Scholar is freely available to use if somebody has internet connection.

6. Searching out literature

Literature searching is the process of finding relevant scholarly documents and databases with a view to using them in the review process. Though, it could be a tedious process for the novice reviewer, as it involves considerable time spent searching manually and performing computer-based searching. But the question remains how literature review should be properly conducted. In this regard, Hek and Helen (2000) suggests a comprehensive rigorous searching approach for accumulating a considerable number of reading materials on which the review will be conducted.

Setting up an appropriate record keeping system, either in paper based or digital database format, can facilitate easy tracking of the record and at the end it will be useful for generating an overall picture of a search history (Timmins and McCabe 2005). Even when searching through the internet, take notes on the online addresses and other information so that it can used later on if necessary. Computer based bibliographic referencing tools, such as Endnote or Reference Manager, are specially designed for managing references. Jesson (2011) presents a number of approaches for searching information, mentioned in what follows:

- Library catalogue
- Digital library/electronic library
- Individual full-text journal databases
- Bibliographic databases
- Official websites
- Online repositories

Most libraries provide the facilities for searching their resources. The library catalogue provides information about the bibliographical details and the locations of all the publications within a library. In most of the developed countries, the catalogues are especially prepared computer databases, which can be accessed on the university network. Electronic searching tools can facilitate easy searching for all kinds of information using keywords. A bibliographical database is an organized list of publications in a particular subject area. On the university library website, it will be possible to search for relevant databases related to the subject area. Some examples of bibliographic database are Web of Knowledge, Medline, ProQuest, etc. Keyword searching is a most common form of identifying literature. The keywords must be very close to the term on which literature is sought. The flexibility of using keywords is that it could be changed with some other close language for a new trial, which could result in different set of literature. Keywords also could be combined with 'Boolean operators' to make variation in search strategy and could generate different results; 'and', 'or' and 'not' are the most commonly used Booleans.

7. Writing literature review

It is mentioned before that a literature review is conducted for receiving a wider understanding about a topic and to get a clear idea about how to proceed with one's own piece of work. However, the most important part of the literature review is to use/apply the information gathered and ideas captured in the process of research undertaking. The application of the knowledge depends on a number of factors, ranging from quality of the reading, to the ability of the researcher to use information for satisfying the readers about the research problem. It requires putting in place a literature review planning and strategy; the strategies may include tactics to summarize what has already been done in the area of research and to do critical evaluations of past works (Carnwell and Daly, 2001). This will help to choose the researcher's arguments and accordingly guide her/him to gather information (facts, expert testimony, examples, explanations) and elements to support the arguments (e.g., anecdotes, visual aids, list of points, rhetorical questions). Moreover, the researcher who is doing a literature review has to be aware of what the reader might be expecting from the review work and this awareness would help the researcher's prior identification of the main points, gather necessary information around those points and develop a well-structured, coherent piece of review work using these contents/information. Finally, the reviewer has to provide narrative illustrations on how she/he has summarized the ideas and information gathered, which will finally demonstrate the researcher's capacity to develop a set of recommendations with good reasoning and make the research results valid.

8. Literature review: geographical relevance in the contexts of Bangladesh

Previous sections describe the necessity of doing a literature review to know about the facts on the topic of concern and for enhancing the existing knowledge base of the researcher. However, conducting a literature review in geography has some intrinsic attributes that may make the review results distinct from the literature review conducted by other disciplinary professionals. In general, the major attributes noticed in the literature review conducted by geographers are of four kinds. These are as follows.

Ascribing locations and contextual settings: Review results conducted by geographers generally adhere to locations, sometimes maps. To compare spatial differences becomes an integral part of the literature review and thus maps, with critical narratives, provide deep insights about the attributes of the phenomenon. These locational comparisons create trends of transformation and help to adopt appropriate hypotheses to test, including the factors that influenced them in the proposed research. For instance, Islam (2006), Sanyal (2006) detected a significant transformation in the forest cover located in the central parts of Bangladesh, popularly known as *Modhupur forest*, where he used satellite images from different times. Reviewing literature relating to this forest cover change may suggest that use of old satellite image data (1962) and could play important roles in constructing the historical distribution pattern of the forests, what no other means could be derived. Rashid (2003) used population census data and converted those data into grid formats compatible with decadal time-series imagery (from 1950 to

2001) and integrated it with local administrative maps. Thus, methodological fusion (combining census data with satellite image) helped him to detect land transformations of Savar (a place near Dhaka) occurring over the last half-century.

- What are the flaws in the policy agenda that create grounds for environmental degradation and deprivation of certain groups of people in the society? How are the policies influenced by power and politics?
- What are the micropolitics and practices of power at the local scale and what are their consequences?
- How are women as a group marginalized due to certain processes of environmental degradation? How have they been expelled from the benefits of social forestry programmes?
- How do certain types of environmental and social degradation cross the boundaries of socio-economic space?
- What are the roles of civil society in the environmental movement in Bangladesh?

Finally, her literature review results indicated that a tendency for profit maximization at an individual level, inefficient roles of state agencies, ineffective policy and institutions arrangements were the factors responsible for environmental degradation.

Understanding ecological integrity: Geographers working in Bangladesh pay attention to understand the interacting actions of physical elements and related processes so as to understand the ecological functions and integrity of an area. Efforts of their literature review are, therefore, aligned with that point of interest. A better understanding of the ecological integrity helps researchers to understand the attributes of constituent ingredients and to measure the degree of change if any in the environmental conditions as a result of the impact of hazards/disturbances (human-made or natural). It is imperative to mention here that an inappropriate and incomplete reading of delta systems (i.e. the forms, process and functions), which might happen as results of inadequate literature review, may mislead the development planning of that region. This took place in Bangladesh, where thousands of kilometres of embankments erected and more than hundreds of polders were built in the coastal deltaic regions with an intention to protect the land from flooding, but finally those projects created serious waterlogging crisis in those areas. P.C. Mahalanobis warned in 1927 that erecting embankments would be problematic for the Bengal delta (Mahalanobis 1927). Satish Chandra Mitra (1916) also indicated the cause and consequences of waterlogging in the regions but these historical documents were not consulted by the United States team (IECO 1964) who proposed creation of polders to solve local floods and cyclone impacts/problems. Mahalanobis (1927) examined time series rainfall data to identify cause and consequences of NorthBengal floods using scientific methods and suggested not to construct physical structures like embankments, roads/railways across the floodplains because it may cause disturbance of water flow in the floodplains. He also suggested combining structural and non-structural measures to combat flood impacts. Mitra (1916) showed how irrigation (for agriculture) is linked with prevalence/spread of malaria in lower deltaic areas. These suggest that knowledge about good and bad use of water resources existed long before in Bangladesh. Taking this knowledge into account while planning and implementing physical structures in the flood and deltaic plains would help to avert many problems that communities of Bangladesh are currently facing.

Understanding scale: Understanding scale is one of the prerequisites to undertaking effective geographical investigations. Different physical, social, economic and ecological elements might appear differently in terms of frequency of occurrence and spatial arrangements with the change of scales. Literature review in this regard could help to know how the elements are behaving at different levels and could help to understand the wider contextual frameworks against which phenomenon occur. The selection of appropriate literary sources would also help to understand the micro conditions and related dynamics of the society.

Understanding historical contexts: The geographer's undertaking of a literature review, in many instances, focuses on understanding the historical contexts of events and occurrences. This kind of reading provides a perspective view/knowledge about the formation or occurrence of physical, social systems and functions. A recently published book by Shamsuddin (2015) entitled *Toponyms of Bangladesh: Footprints of History* is a significant contribution in this regard. It shows how the Bangla language, along with developments in allied areas such as social, religious, economic and political processes, has evolved in this deltaic regions through ahistorical progression of actions and events. Place names of Bangladesh have been used by the author as a proxy to show how the fusion of multiple variables of external and internal origins contributed to the evolution of the Bengali language and also facilitated the process of emergence of the country as a sovereign nation. In addition to the historical episodes and chronicles presented, Shamsuddin (2015) also illustrates the influence of local physical attributes/processes in labelling a place with a certain name.

The place names of Bangladesh are thoroughly analyzed in the book for their historical lineage, structural forms and patterns and literary sources. A spatial analysis, using a Geographic Information System (GIS), is performed and presented with geographical clustering of different prefixes and suffixes of place names and then superimposed on the physical maps of Bangladesh. These maps give indications that the place names have a strong correlation with their being with physical and social phenomena (like distribution of canal, rivers, sediment deposited islands, presence of settlements) of an area.

9. Concluding remarks

Thrift (2002) in a well cited paper indicated both the opportunities and challenges of the discipline of geography. He mentions that new factors are coming into appearance and influencing and shaping out disciplinary outfits; new forces like large-scale computing facilities are helping to create strong visual reality simulations, web-based surveys, space-time vectors, contributions to political agenda, and will put geographers in more challenging conditions to align, modify existing tools and methods to accommodate these dimensions. In Bangladesh. vis-à-vis those working in the periphery, geographers will be in more trouble where applying theoretical frameworks in investigations is already inadequate. Strong and effective literature reviews may play important roles in this regard by informing the researchers about the works performed by geographers through innovating the use of tools and methods. Literature review may also inform about the contingent settings of the study area that determine the application strategies for the proposed research.

References

Baumeister, R. F. and M. R. Leary. 1997.'Writing Narrative Literature Reviews'. *Review of General Psychology*, 1(3): 311–320.

Burns, N. and S. Grove. 1999. *Understanding Nursing Research*, 2nd ed. Philadelphia: W. B. Saunders Company.

Carnwell, R. 1997. 'Critiquing Research'. *Practice Nursing*, 8(12): 16–21.

Carnwell, R. and W. Daly. 2001. 'Strategies for the Construction of a Critical Review of the Literature'. *Nurse Education in Practice*, 1: 57–63.

Coughlan, M., P. Cronin and F. Ryan. 2007. 'Step-by-Step Guide to Critiquing Research. Part 1: Quantitative Research'. *British Journal of Nursing*, 16: 658–663.

Cronin, P., F. Ryan and M. Coughlan. 2008. 'Undertaking a Literature Review: A Step-by-Step Approach'. *British Journal of Nursing*, 17(1): 38–43.

Garrard, J. 2014. *Health Sciences Literature Review Made Easy*. New York: Jones and Bartlett Learning.

Green, B. N., C. D. Johnson and A. Adams. 2006. 'Writing Narrative Literature Reviews for Peer-Reviewed Journals: Secrets of the Trade'. *Journal of Chiropractic Medicine*, 5(3): 101–117.
Hart, C. 1998. *Doing a Literature Review*. London: Sage Publications.
Hek, G. and L. Helen. 2000.'Systematically Searching and Reviewing Literature'. *Nurse Researcher*, 7(3): 40–53.
IECO (International Engineering Company Inc.). 1964. *Master Plan: East Pakistan Water and Power Development Authority*. California.
Islam, S. T. 2006. *Resources Assessment of Deciduous Forests in Bangladesh*. PhD Thesis, Durham University, Durham, UK.
Jesson, J. 2011. *Doing Your Literature Review: Traditional and Systematic Techniques*. London: Sage Publication.
Knopf, J. W. 2006. 'Doing a Literature Review'. *Political Science and Politics*, 39(1): 127–132.
Leedy, P. D. and J. E. Ormrod, 2005. *Practical Research*. Upper Saddle River: Pearson Custom.
Levy, Y. and T. J. Ellis. 2006.'A Systems Approach to Conduct an Effective Literature Review in Support of Information Systems Research'. *Informing Science Journal*, 9: 181–212.
Mahalanobis, P. C. 1927. *Report on Rainfall and Floods in North Bengal, 1870–1922*. Calcutta: Bengal Secretariat Book Depot.
McCabe, S. 2005. 'Who Is a Tourist? A Critical Review'. *Tourist Studies*, 5(1): 85–106.
Mitra, S. 1916. *History of Jessore and Khulna* (in Bengali). Dhaka: Lekhok Samobay.
Onwuegbuzie, A. J. and F. Rebecca. 2016. *Seven Steps to a Comprehensive Literature Review: A Multimodal and Cultural Approach*. Los Angeles, CA: SAGE.
Peet, R. 1998. *Modern Geographical Thought*. London: Blackwell Publishers.
Petticrew, M. 2001. 'Systematic Reviews From Astronomy to Zoology: Myths and Misconceptions'. *BMJ*, 322: 98.
Rashid, M. S. 2003. *A Study of Land Transformation in Savar Upazila, Bangladesh, 1951–2001: An Integrated Approach Using Remote Sensing, Census, Map and Field Data*. PhD Thesis, Durham University, Durham, UK.
Ridley, D. 2012. *The Literature Review: A Step-by-Step Guide for Students*, 2nd ed. Los Angeles: SAGE.
Sanyal, N. 2006. *Political Ecology of Environmental Crises in Bangladesh*. Durham Theses, Durham University. http://etheses.dur.ac.uk/2893/.
Shamsuddin, S. D. 2015. *Toponyms of Bangladesh: Footprints of History* (in Bengali). Dhaka: Sahitya Prakash.
Thrift, T. 2002.'The Future of Geography'. *Geoforum*, 33(3): 291–298.
Timmins, F. and C. McCabe. 2005. 'How to Conduct an Effective Literature Review'. *Nurs Stand*, 20(11): 41–47.
Webster, J. and R. T. Watson. 2002. 'Analyzing the Past to Prepare for the Future: Writing a Literature Review'. *MIS Quarterly*, 26(2): 13–23.
Younger, P. 2004. 'Using the Internet to Conduct a Literature Search'. *Nursing Standard*, 19(6): 45–51.

15 Local trends, the current state and the future of geography

Alak Paul and Sheikh Tawhidul Islam

1. Introduction

The theme, scope and dimensions of geographical exercises, both in research and teaching activities, at global and local levels, have changed their direction over time. The pioneering stage of geography in Bangladesh has modified its own fashion over the decades, mainly due to the gradual increase of internal diversity of the discipline. Across time, major branches of geography divided into many sub-fields and also absorbed many new sub-fields. New tools and techniques are incorporated into the domain of geographic investigations and provide useful results and deep insights. Some old branches become ostracized by geographers. For instance, sub-fields of human geography like rural geography, agricultural geography, industrial geography, historical geography were popular among the geographers in Bangladesh at the early stage, but gradually these (sub-fields) have lost popularity among the researchers as well as tertiary level students because of their irrelevance for the time and the appearance of more interesting research frontiers. In contrast, some branches of human geography, like urban geography, population geography and economic geography, are becoming popular among the geography academicians and professionals. It is imperative to mention that most of the studies tend to show areal differentiation in the occurrence of phenomenon and locational patterns of systems and processes in their respective fields. In some attempts, geographers focused on physical geography, especially in the fields of geomorphology, quaternary environments, disaster management and planning. In addition, ecology becomes an important component in the geography syllabus in all public universities of Bangladesh, which contributes in enhancing the capacity and encouraging postgraduate researchers to conduct the ecology focused study.

2. Geographical methods used by Bangladesh geographers

Geographers in Bangladesh deploy a range of different tools and techniques in their geographical investigations, but there is a biased use of quantitative approach over the qualitative methods. The quantitative methods were mainly used in the research works carried out in earlier times due to the influence of the quantitative revolution in the West. In the quantitative approach, the questionnaire survey appeared to be the popular data collection technique in Bangladesh and statistical tools like central tendency, multivariate analysis, correlation and regression, inference drawing, significance testing were used for data analysis purposes; applications of these quantitative techniques still remain in prominence with geographers. In recent times, qualitative methods are getting popular among the geographers. Return of young academic staffs after receiving training from Western universities, mainly through PhD programmes and the societal need for micro-geographic investigations, which need qualitative methods (like in-depth interview, observation, group discussion, participatory appraisal processes), influence Bangladesh geographers to apply various qualitative methods in their research exercises.

Spatio-temporal analysis is also getting popular nowadays among geography professionals in Bangladesh through the application of Geographical Information Systems (GIS), Remote Sensing (RS) and Global Positioning Systems (GPS) techniques. The capabilities of these techniques to handle large databases (both spatial and non-spatial; vector and raster), the ability to apply various spatial analysis techniques like overlay, union, intersect, geo-statistics have contributed to making geographers interested in acquiring skills in these methods. The skills enhancements of geographers and related contributions made many government and non-government agencies interested in recruiting geographers in important professional positions. However, one of the limitations remaining in geographical investigations in Bangladesh is a lack of epistemological and ontological enquiring in research exercises. Most geographers in Bangladesh ignore or pay very little attention to applying theories in their research activities. Only few research works where qualitative methodologies are applied show the tendency to apply theories so as to fit their results in a broader conceptual framework.

3. Major contribution of the discipline 1

3.1 Cater to the human resource needs of the country

The geography departments in the universities of Bangladesh have been playing important roles in producing graduates who are serving the nation where geographical knowledge is necessary. Teaching in the schools, colleges and universities is one of the major areas where the geography graduates are absorbed. At present, about 140,000 government and registered non-government educational institutions are teaching about 30.5 million students in different subjects by engaging about 929,000 teachers in Bangladesh. Currently, 38 NCTB (National Curriculum and Textbook Board) books in different classes (Class I to XII) contain geographical elements to teach the students. In addition, a number of subjects, such as Bangladesh and Global Studies, Geography, Agriculture, Science encompass geography and environment and related aspects, including climate change and disaster management issues. Apart from the teaching professions, geographers are also contributing in different agencies working in environmental pollution, urban and regional planning, disaster management and climate change aspects. The highly trained geography professionals (especially those who have PhD degrees) are working as frontline researchers in filling up knowledge gaps through conducting high quality research exercises.

4. Major contributions of the discipline 2

4.1 Geographers' contributions in filling out knowledge gaps

Geography is one of the old disciplines in Bangladesh; at Dhaka University, the department was established in 1947, at Jahangirnagar University, in 1969 and at Rajshahi University, in 1955. Apart from producing graduates, the geography departments have also been playing vital roles in conducting research activities. All the departments run MPhil and PhD programmes where students from different quarters enrol to conduct research exercises. In addition, faculty members of geography departments regularly do research works for different government and non-government agencies of the country. In some of the areas, geographers play the lead roles in the society to fill in knowledge gaps. The works on migration of people (Elahi etal. 1999), development induced displacement (Khatun 2005), gender geography (Ahsan etal. 2005), quaternary geography, environmental impact assessment (Shamsuddin etal. 2009), urbanization process, land and man interface in dynamic riverine islands of Bangladesh are some examples of works conducted by Bangladesh's eminent geographers. These works bring many new insights to different problem domains and help policy planners make appropriate decisions.

5. Concluding remarks and the future of geography

The discipline geography has never been static; rather it has evolved with the progress of the society in terms of its growth and development (Koutsopoulos 2011). But aspects such as the changes in the society by means of development and resulting historical-material (Harvey 2004) progression were not properly scrutinized by commissioning research projects, at least in this part of the world. The research works presented in different chapters of this book are, in most of the cases, conducted by personal motivation and initiatives, not conducted through institutional incentives. The lack of institutional encouragement in conducting research results in fewer professionals involved in research activities and hence limited contributions in developing knowledge products. This inadequate effort in producing geographical knowledge and explanations based on empirical findings causes the geographers of the Global South not to be in line with the evolution experienced by the geographers working in the Global North. But whatever contributions the southern geographers are able to make, at least, are able to make a strong argument that more robust efforts are necessary to generate contextualized knowledge based on local social and physical micro-elements, which might be culminated through cumulative observations and examinations. Against this backdrop, discussions in regards to the future trends of geographical works for both South and North are provided in the following sections.

In recent times, the opportunities to handle and analyze big volume data has created a fresh scope to the way geographers think and the way they should model the functions and processes of the societies and the world; geographers can now visualize the historical changes (e.g., time-lapse visualizations of thematic sectors like land use, forest cover change of an area) in association with real time conditions of that sector. It is imperative to mention here that (a) the advancements of technology, especially in computational processing and analysis, (b) cyber power and its penetration deep into the society, especially by means of social media facilities including the use of Internet of Things (IOT) to connect machines with humans, (c) easy application of location based data gathered by satellite sensors during the last 40 years (e.g., Landsat data by using Google Earth Engine facilities for researchers), in connection with the non-satellite data like elevation, topography, hydro-meteorological variables, have collectively created a condition to read and characterize the societies differently. This condition has persuaded geographers to define geographic space differently, which is 'less fragmented' compared to the past views. Looking at the societal changes with the aid of the new technologies and methods may help to depict the society as 'whole of elements' and geography, here, both in the South and North, may find ways to redefine its epistemological makeup to examine the things and phenomena. The ontological framework may also emerge accordingly with necessary methods which will be 'integrated' rather than 'separated'. The time when different disciplines are making their boundaries strong to keep them pure, geography at that time has got the opportunity to open it up, to be broader rather than confined into fragmented interpretations relying only on data/information. Harvey's suggestions in this regard may be deemed appropriate to create a new wave of geographical revolution in terms of taking an interdisciplinary (integrated) approach in defining and characterizing the society. His propositions include that dialectic analysis is required to better understand the historical-geographical materialism of a society in a holistic way that finally helps to apprehend the values of the societies through which people interact with micro-elements of the society and make progress (Harvey 2004). Citation of one example might be appropriate here: Rigg (2006) shows that the diffusion of technology has changed the social-economic-cultural landscape of rural Thailand. Further interpretations of this technological penetration into rural settings in Thailand might help the geographers to examine the changes of characteristics between 'current rural' compared to 'past rural' in Thailand; similar conditions and changes also exist in Bangladesh rural settings. Harvey's (2004) suggestion of mode of explanation with the aid of technological potency, as indicated earlier, might help here to better understand the geographical space of people and society for whom geographers work.

Geographers need to work beyond the core-periphery debate. They need to commission joint exercises that will eventually build bridges between them and contribute in making models with the capacity to accommodate the vital macro and micro elements of social, physical and environmental functions and processes. In doing so, both the empirical exercises and theory-based investigations will team up and will eventually reinforce the discipline. In other words, it will be difficult to develop theories and models for dynamic societies without having strong collaborative projects and performing related activities, such as holding colloquia and conferences on the issues and subjects in concern for critical analysis of data/information derived. Relevant international geographical associations, institutions, societies and academic departments may work together with Southern partners to fill in the gaps in capacity and knowledge domains. They will make the opportunity real, to appear as a cutting-edge discipline that is integrated in approach, elastic at scale, necessary to solve contemporary, multifaceted problems at both the local and global levels.

References

Ahsan, R. M., N. Ahmad and H. Khatun. 2005. *Gender Geography; A Reader: Bangladesh Perspective*. Dhaka: Bangladesh Geographical Society.
Elahi, K. M., K. S. Ahmed and M. Mafizuddin (eds.). 1999. *Riverbank Erosion, Flood and Population Displacement in Bangladesh*. Dhaka: REIS-JU-UM.
Harvey, D. 2004. *Justice Nature and the Geography of Difference*. Singapore: Blackwell Publishing.
Khatun, H. 2005. 'Development-induced Displacement and Rehabilitation in Bangladesh: Challenges and Perspectives'. *Oriental Geographer*, 49(2): 135–150.
Koutsopoulos, K. C. 2011. 'Changing Paradigms of Geography'. *European Journal of Geography*, 1: 54–75.
Rigg, J. 2006. 'Land Farming Livelihoods and Poverty: Rethinking the Links in the Rural South'. *World Development*, 34(1): 180–202.
Shamsuddin, D. S., S. T. Islam, M. S. Rashid and M. S. Alam. 2009. *National Adaptation Interventions to Address Climate Change Induced Flood Hazard: A Strategic Environmental Assessment*. Unpublished Report conducted for ActionAid Bangladesh, Dhaka.

Index

Note: figures are denoted with *italic* page numbers; tables are denoted with **bold** page numbers; note information is denoted with n and note number following the page number.

AAG (Association of American Geographers) 45
academia: feminism in 30–31; *see also* education; research and scholarship
Acquisition and Requisition of Immovable Property Ordinance 1982 II (ARIPO) 186, 188, 192, 198
Action Theory 207
administration: decentralization of 12; military 12; urban concentration of 12, 14; wetlands 230
agriculture: British period decline of 34; climate change crisis impacts on 147, 153–154, 156–160, **159**, *159*; environmental assessments of **128**; food security with domestic 167, 169, *169*, 171–172, **174**, 180; land for 21, 74–93, 172–173, 180, 222, 224, 226–227, 229; scarcity and landlessness for workers in 2–3, 74–93, 157–158; sustainable 167, 179, 180, 180n1; urban land for 21
Ahmad, Nafis 15
Ahmad, Nasreen 2–3, 35, 36, 37, 38, 74
AHP (Analytic Hierarchy Process) 213, 215
Ahsan, E. 34
Ahsan, Rosie Majid 2, 18, 26, 34, 37, 38, 48

air: environmental assessments of 124, **129**, 132, **137–142**; pollution 13, 27
Alam, Shamsul 49
Aligarh Muslim University 48
Analytic Hierarchy Process (AHP) 213, 215
animals *see* agriculture; flora and fauna
applied geography: disaster and vulnerability research (*see* disaster and vulnerability research); displacement and resettlement 4, 49–50, 185–200; food security 4, 83, 92, 161, 167–181; literature reviews 4, 241–252; local trends, current state, and future of geography 4, 254–258; natural resource appraisal 4, 222–239; overview of 3–4; research and scholarship in 5–6
Arefin, Sirajul 49
ARIPO (Acquisition and Requisition of Immovable Property Ordinance 1982 II) 186, 188, 192, 198
Asian Development Bank 127, 185
Association of American Geographers (AAG) 45
Atkins, Peter J. 64

Banaras Hindu University 48
Bangladesh Agricultural Research Council (BARC) 153, 175, 227

Index 261

Bangladesh Bureau of Statistics (BBS) 75, 151, 156, 157, 173, 175, 179
Bangladesh Geographical Society (BGS) 35, 36, 37
Bangladesh India Electrical Grid Interconnection Project (BIEGIP) 185
Bangladesh Meteorological Department (BMD) 235
Bangladesh Power Development Board (BPDB) 235
Bangladesh Rural Advancement Committee (BRAC) 84, 198
Bangladesh University of Engineering and Technology 15
BARC (Bangladesh Agricultural Research Council) 153, 175, 227
BBP (Bhairab Bridge Project) 185, 191, 192
BBS (Bangladesh Bureau of Statistics) 75, 151, 156, 157, 173, 175, 179
Beaujeu- Garnier, J. 45
Begum Rokeya University 49
BGS (Bangladesh Geographical Society) 35, 36, 37
Bhairab Bridge Project (BBP) 185, 191, 192
BIEGIP (Bangladesh India Electrical Grid Interconnection Project) 185
biomass energy 232, 234, **234**, 235–236
BMD (Bangladesh Meteorological Department) 235
Bondi, Liz 30
Boserup, Ester 30
BPDB (Bangladesh Power Development Board) 235
BRAC (Bangladesh Rural Advancement Committee) 84, 198
Brundtland Report 181n3
Buff, Eva 30
business and industry: development programmes stressing 28; energy consumption by 233; environmental impact assessments for 124, 127, **131**; environmental impacts of 27–28; research and scholarship on 19; urban concentration of 12–13, 14, 19

Canadian International Development Agency (CIDA) 41
CAPE (Convective Available Potential Energy) 152
CARE (Cooperative for Assistance and Relief Everywhere) 41
CEIP-I (Coastal Embankment Improvement Project) 185
Centre for Urban Studies (CUS) 11, 17–18
Chapman, Murray 35
Chatterjee, Leena 35
children: displacement and resettlement of 199; education for (*see* education); food insecurity for 170, 171; human trafficking 34; urban context for 21; women's responsibility for 28, 39, 160
Chittagong: disaster and vulnerability research in 215; palaeogeography and palaeoenvironment of 115; urbanization and growth of 12, 13, 21
Chittagong Hill Tract 112, 119, 176
Chittagong Hill Tracts Rural Development Project (CHTRDP) 185
Chittagong University 38, 49
Chittagong Uria Fertilizer Ltd. 127
cholera 44
Chowdhury, Mohammad Abu Taiyeb 3–4, 167
Chowdhury, Tanwir Adib 180
CHTRDP (Chittagong Hill Tracts Rural Development Project) 185
CIAs (Cumulative Impact Assessments) 126, 144
CIDA (Canadian International Development Agency) 41
cities *see* urbanization
Clarke, John I. 46–47, 47
climate: natural resource appraisal of 227; palaeo- 97, 98, 99, 100–101, 105, 112, 117–118; *see also specific types of weather events*
climate change crisis: climate variables in Bangladesh 147–151, **148–150**; cyclones with 147, 151, 152; disasters and vulnerabilities

262 *Index*

due to 147, 150–156, *152*, **155–156**, 160–161, 206, 211; droughts with 151, 153; economic impacts of 151, 156, **156**; environmental impact assessment for 135–144, **137–142**, *143*; floods with 147, 151; food security affected by 172; gender dimensions of 158–160, **159**, *159*; global *vs.* local perspective on 146; government policies and programs on 146–147, 161; impacts of, in Bangladesh 151–156, *152*, **155–156**; monsoon rainfall with 150, **150**, 153, 160–161; overview of 3, 146–147, 160–161; population geography study of 49–50; poverty nexus with 156–158, *157*; salinity intrusion with 151, 153–154, **155**; sea level change with 151, 155, 172; slow and progressive hazards with 151, 153–156, **155–156**; social dimensions of 156–160, *157*, 159, *159*; spatial and temporal context for 146–147, 151–156; sudden big events with 151–152, *152*; temperature change and 147–150, **148–149**, 151, 161; thunderstorms and lightning hazards with 151–152
Climate Vulnerability Index 215
coal, oil, and natural gas 232–234
Coastal Embankment Improvement Project (CEIP-I) 185
Coastal Vulnerability Index (CVI) 214–215
Coleman, Alice 225
commerce *see* business and industry; trade
communication: displacement and resettlement plan 191; infrastructure lack 16
Conference on Commercial Activities, Women and Ecology 36–37
Convective Available Potential Energy (CAPE) 152
Cooperative for Assistance and Relief Everywhere (CARE) 41

core geography 1, 4–5, 6, 257
credit *see* loans and credit
crimes 20, 64, 89, 90
Crisis and Conflict Theory 207
cultural context: cultural geography studying 27; displacement, resettlement, and awareness of 197, 200; gender in 26 (*see also* gender geography); HIV/AIDS geography in 57–58, 60–61, 62–64; scarcity and landlessness influenced by 92; technology changes to 257; urban 12
Cumulative Impact Assessments (CIAs) 126, 144
current state of geography *see* local trends, current state, and future of geography
CUS (Centre for Urban Studies) 11, 17–18
CVI (Coastal Vulnerability Index) 214–215
cyclones: climate change crisis and 147, 151, *152*; disaster and vulnerability research related to 214, 215; food insecurity with 171; scarcity, landlessness, and vulnerability to 75, 82, 92; women impacted by 34

dating, palaeogeographic 108–109, **110–111**
Decade for Women 28, 30
DEEP (Dhaka Elevated Expressway Project) 186
deep sea cores 99–100
Department for International Development (DFID) 185
Department of Environment (DoE) 127, **131**
development programmes: displacement and resettlement induced by 4, 185–200; environmental impact assessments for 124–144; food security 40–41, 167, 179–180; gender geography and 27–30, 39–41; literature reviews on effects of 250–251; population geography linkage with 50–51; public-private

partnerships funding 186, 198–199; rural 167, 179–180, 180–181n2

Dey, N. K. 37

DFID (Department for International Development) 185

Dhaka: floods in 19–20; land access and distribution in 19; palaeogeography and palaeoenvironment of 115; rural-urban migration to 17; scarcity and landlessness in 76; slums and squatter settlements in 16–17, 18; solar energy in 236–237, **236–237**; urbanization and primacy of 12–13, 16, 21; wetlands in and around 229

Dhaka Elevated Expressway Project (DEEP) 186

Dhaka University: gender geography research and scholarship at 35–36, 37–38; geography department at 256; population geography research and scholarship at 48; urbanization research and scholarship at 15

diatom analysis 106–108, *107*

Disaster and the Silent Gender: Contemporary Studies in Geography (Ahsan & Khatun) 37

disaster and vulnerability research: climate change crisis and 147, 150–156, *152*, **155–156**, 160–161, 206, 211; Disaster Resilience of Place model in 209–210; displacement and resettlement in 197–198, 199, 200; double structure of vulnerability model in 206–207, 210; on economic impacts of disasters 151, 156, **156**; exposure analysis in 211–212; food insecurity in 170, 171, 172, **174**, 177; gender geography in 34, 37; global environmental change model in 207–208, 210; holistic approach in 209, 210; mapping in 210–216; methodologies and models in 202, 205–216; multi-criteria approach in 210–216; natural hazard analysis in 211; overview of 4, 202, 216; preparedness analysis in 212; Pressure and Release model in 208–209, 210; prevention analysis in 212; response analysis in 212; risk and hazard model in 207, 210; risk consideration in 202, **203–205**, 205–210; scarcity and landlessness in 75, 76–77, **78–79**, 82–83, 89, 90, 92, 157–158; socio-economic variables in 206, 208–210, 211, 213–215; spatial and temporal context for 208, 209–210, 211, 213–216; vulnerability analysis in 212–213; vulnerability definitions and terminology in 202, **203–205**, 210, 216

Disaster Resilience of Place (DROP) model 209–210

displacement and resettlement: communication and consultation on 191; compensation for 188–190, **189**, 192, **193–195**, 196–200; current resettlement practices 190; employment effects 186, 188–190, 196, 197, 198, 199; good practices for 190–196, **193–195**, 198; identification cards with 191; improvements to 199–200; infrastructure and 185–186, 191–192, 196–197; Jamuna Multipurpose Bridge Project inducing 4, 185, 187, 190–192, 196–198; legal framework of 186, 187–188, 198–199, 200; minimization of 191; overview of 4, 185–187, 198–200; population geography study of 49–50; poverty with 186–187, 190, 200; requirements for resettlement 188–190, **189**; resettlement and rehabilitation planning with 186–187, 190, 192, 196, 198; resettlement process 196–197; resettlement sites 197, 200; risk mitigation with 186–187; social changes with 186, 192, 196, 197, 199, 200; videofilming before 191

divorce, women's rights in 33

DoE (Department of Environment) 127, **131**
Domash, Mona 30
double structure of vulnerability model 206–207, 210
DROP (Disaster Resilience of Place) model 209–210
droughts: climate change crisis and 151, 153; food insecurity with 171, 176; scarcity, landlessness, and vulnerability to 82, 92
drug users, HIV/AIDS among 59, 60, 61, 62
Duncan, O. D. 46
Dunn, Christine E. 64

earthquakes 118, 151, 176, 214
Earth Resources Technology Satellite (ERTS) 226
East Bengal Requisition of Property Act (1948) 187–188
An Economic Geography of East Pakistan (Ahmad) 15
economy: climate change disasters' effects on 151, 156, **156**; development programmes focus on 28; economic geography 27; gross domestic product 13; informal sector of 19, 20–21; socialist 16; urban 12, 14, 19, 21; War of Liberation damages to 16; *see also* business and industry; socio-economic status; trade
education: climate change impacts on 146; development programmes to improve 28; disasters and vulnerability with lack of 215; feminism in 30–31; food insecurity in relation to 170, **174**, 177; in geography 255–256; research and scholarship on 20; scarcity and money for 86, 92; urban concentration of 12, 14, 20; women's 33; *see also* research and scholarship
EIAs *see* environmental impact assessments
Elahi, K. Maudood 2, 44, 47, 48, 49
Elahi, Moudud 38
EMAP (Environmental Management Action Plan) 198

employment: agricultural 2–3, 74–93, 157–160, **159**, *159*, 198; displacement and effects on 186, 188–190, 196, 197, 198, 199; food security affected by **174**; informal sector 19, 20–21; natural resource protection in conflict with 230; scarcity relative to 82–83; urban 19, 20–21; women's 20–21, 29, 34, 39–40, 41, 158–160, **159**, *159*
EMPs (Environmental Management Plans) 124, 132
energy: Convective Available Potential Energy 152; environmental assessments related to **128**; infrastructure for 20, 232; natural resource appraisal of 232–237, **234**, **236–237**; non-renewable (oil, gas, coal) 232–234; renewable (biomass, hydropower, solar, tidal, wind) 232, **234**, 234–237, **236–237**; urban 20
environmental concerns: climate change crisis 3, 49–50, 135–144, 146–161, 172, 206, 211; development programmes and 28–29, 124–144; disaster and vulnerability research on 206, 207–208, 211, 215 (*see also under* climate change crisis); floods as (*see* floods); food security affected by 172; industrial impacts on 27–28; literature reviews for understanding 250–251; natural resource appraisal due to 4, 222–239; research and scholarship on 19–20; urbanization effects on 13, 19–20; women addressing 28–29; *see also* environmental impact assessments
Environmental Conservation Act (2000/2001) 125
Environmental Court Act (2000) 125
environmental impact assessments (EIAs): of air effects 124, **129**, 132, **137–142**; case studies of 131–144; of flora and fauna effects 124, 127, **130**, 132, **137–142**; of land effects 124,

128, 132, **137–142**; legislative frameworks and regulations on 127, **131**; for NORI 131–135, **133–134**; overview of 3, 124–125; scope and definition of 125–127, **128–130**; of social factors 124, 126, 132, **137–142**; Strategic Environmental Assessments 126, 135–144, **137–142**, *143*; of water effects 124, 125, **129**, 132, **137–142**
Environmental Management Action Plan (EMAP) 198
Environmental Management Plans (EMPs) 124, 132
Environment Conservation Act (1995) 127
Environment Conservation Rules (1997) 125, 127
Environment Policy (1994) 125, 127
equality, gender 29–30, 31
Eratosthenes 26
erosion *see* riverbank erosion
ERTS (Earth Resources Technology Satellite) 226
Eusuf, Ammat-uz-Zohra 35, 37
exposure analysis 211–212

FADEP (Integrated Food Assisted Development Programme) 40
families: kinship networks influencing rural-urban migration 17; women's roles in 28; *see also* children
famines: climate change crisis and 147, 161; decline of 167; food insecurity with 170, 171; rural-urban migration due to 17; women impacted by 34
FAPs (Flood Action Plans) 125
Female Migrants' Adaptation in Dhaka: A Case of the Process of Urban Socio-Economic Change (Huq-Hussain) 37
feminism 29–32; *see also* gender geography
financial issues: displacement and resettlement-related 188–190, **189**, 192, **193–195**, 196–200; loans and credit as 29, 83–86, **85**, 170, 192, 196, 197, 199; urban housing 18; *see also* economy; employment; income; poverty; socio-economic status
fires 103, 214
FIVIMS (Food Insecurity and Vulnerability Information Mapping Systems) 168, 178–179
Flood Action Plans (FAPs) 125
Flood Plan Coordination Organization (FPCO) 125
floods: climate change crisis and 147, 151; disaster and vulnerability research related to 213, 214; displacement and resettlement due to 198; environmental impact assessment for 125, 135–144, **137–142**, *143*; food insecurity with 171, 172; literature reviews of development programmes addressing 250–251; research and scholarship on 19–20; rural-urban migration due to 17; scarcity, landlessness, and vulnerability to 75, 76–77, **78–79**, 82, 90, 92; women impacted by 34
flora and fauna: biodiversity of 27, 124, 132; environmental assessments of 124, 127, **130**, 132, **137–142**; natural resource appraisal of 4, 222–239; palaeo- 98, 99, 105–106, 109, 115–117, *116*; *see also* agriculture
Food Insecurity and Vulnerability Information Mapping Systems (FIVIMS) 168, 178–179
food security: agriculture and 167, 169, *169*, 171–172, **174**, 180; conceptual frameworks for 168–171, *169*; definition of 168; disasters and vulnerability of 170, 171, 172, **174**, 177; displacement and loss of 186; food accessibility and *169*, 169–170, 172–173, **174**, 177; food availability and 169, *169*, **174**, 177; food insecurity *vs.* 170–171; food utilization and *169*, 169–170, 172, **174**, 177; further readings and resources on 178; gender dimension of 40–41,

170, 171; government policies and programs for 40–41, 161, 167, 173–175, 178–179; health dimension of 170, 171, 172–173, **174**, 177; key indicators of insecurity 173–174, **174**; mapping insecurity 168, 173–177, 178–179; national state of 171–173; overview of 4, 167–168; poverty and 167, 170–171, 172–173, **174**, 177, 179; regional patterns of insecurity 175–176; research methodology to study 173–177, **174**; research suggestions to study 177–180; scarcity, landlessness, and lack of 83, 92, 173; social context for 172–173; sustainable development context for 4, 167–168, 171–173, 178–180; technological issues affecting 171–172; temporal dimension of 171
Food Security Atlas of Bangladesh 173, 174, 176–177, 178–179
Food Security for Ultra Poor (FSUP) 41
Food Security for Vulnerable Group Development (FSVGD) 40–41
forests: food insecurity in areas of 176; literature reviews of research on 249–250; natural resource appraisal of 222, 224, **231**, 231–232; palaeo- 98, 103, 106, 108, 115–117, *116*, 120; sustainable development approach to 179, 180
FPCO (Flood Plan Coordination Organization) 125
FSUP (Food Security for Ultra Poor) 41
FSVGD (Food Security for Vulnerable Group Development) 40–41
future of geography 257–259

GAD (Gender and Development) 30
Ganges-Brahmaputra-Meghna (GBM) delta 112–113, 154
Garnier, Beajeu, J. 27
gay, lesbian, bisexual individuals 31, 57

GB (Grameen Bank) 84
GBM (Ganges-Brahmaputra-Meghna) delta 112–113, 154
GDP (gross domestic product) 13
Gender and Development (GAD) 30
Gender Development and Environment course 38
gender geography: in academia 30–31; background of 26–27; in Bangladesh 33–39; climate change crisis nexus with 158–160, *159*, *159*; development programmes and awareness of 27–30, 39–41; employment issues in 29, 34, 39–40, 41, 158–160, **159**, *159*; equality of sexes in 29–30, 31; feminist advocacy and epistemology in 29–32; food security/insecurity and 40–41, 170, 171; HIV/AIDS and 58, 59–60; methodologies in 32–33; overview of 2, 26, 41; poverty and socio-economic status in 28, 29, 31; religion and 33–34; research and scholarship in 34–39; social context for 27, 29–31, 32, 37, 39–40, 41; spatial context for 27, 32–33; *see also* women
Gender Place and Culture 30
Geographical Information System (GIS): current use of 255; in disaster and vulnerability research 210, 211, 213, 214, 216; in environmental impact assessments 132, 135, 144; in food insecurity mapping 175; in HIV/AIDS geography 57; literature reviews of data from 252; in natural resource appraisal 227–228, 231, 238; in population geography 49, 50, 51, 52
Geographically Weighted Regression (GWR) 57
geography in Bangladesh 1–6, 254–258; applied (*see* applied geography); human (*see* human geography); physical (*see* physical geography)
Geography of Gender and Human Development course 38

Index 267

GHI (Global Hunger Index) 171
GIS *see* Geographical Information System
Global Hunger Index (GHI) 171
Global North geography 1, 4–5, 6, 257
Global Positioning Systems (GPS) 131, 144, 255
Global South geography 1, 4–6, 257–258; *see also* geography in Bangladesh
Google/Google Scholar 247
government: administration 12, 14, 230; climate change policies and programs 146–147, 161; coordination lack in 19; development programmes (*see* development programmes); environmental concerns addressed by 19–20, 125, 127, **131**; food security policies and programs 40–41, 161, 167, 173–175, 178–179; HIV/AIDS policy and practice 62–63; mismanagement and corruption of 19, 230, 231; urban poverty programmes 17–18
GPS (Global Positioning Systems) 131, 144, 255
Grameen Bank (GB) 84
gross domestic product (GDP) 13
GWR (Geographically Weighted Regression) 57

Hagerstrand, T. 27, 35
Harrison-Church, R. J. 47
Hasan, Zahid 49
Haslett, Steve 180
Hauser, P. 46
HDI (Human Development Index) 171
health geography: cholera 44; development programmes to improve 28; epidemiologic methods in 44; food security/insecurity and 170, 171, 172–173, **174**, 177; gender research and 37, 58, 59–60; HIV/AIDS 2, 54–64; medical geography *vs.* 54; population geography influenced by 44, 49, 50; research and scholarship on 20; scarcity in relation to 82, 83, 86, 89, 92; urban concentration of 12, 14, 20
HIV/AIDS geography: in Bangladesh 59–63; cultural context for 57–58, 60–61, 62–64; gender and 58, 59–60; identity and stigma in 58, 59, 60, 61; lifeworlds' issues in 60–61; mapping of 55, 57, 63; migration and 56–57; overview of 2, 54–55, 63–64; place concern in 62; policy and practice with 62–63; political context for 54, 57–58, 60, 62–63; qualitative focus on 57–58, 63–64; quantitative focus on 55–57, 63; research and scholarship in 55–64; risk behavior and coping in 56, 58, 61–62, 63; social context for 54–64
Hollema, Siemon 180
housing: disasters and vulnerability of 215; displacement and changes to 186, 191, 196; environmental impact assessments for 124; landlessness and lack of 76; private sector 18; public 18; research and scholarship on 18–19; slums and squatter settlements 16–17, 18, 21, 76, 197, 199, 200; solar energy with 236–237, **236–237**
Human Development Index (HDI) 171
human geography: gender-based 2, 26–41, 58, 59–60, 158–160, 170, 171; HIV/AIDS-related 2, 54–64; overview of 2–3; population 2, 27, 44–52; research and scholarship in 5–6; scarcity and landlessness 2–3, 74–93, 157–158, 173; urbanization 2, 11–21, 173
human trafficking 34, 62
Huq-Hussain, Shanaj 17, 18, 34, 37, 38
Hutton, James 112
hydropower 232, 234, **234**, 235

ice cores 99
IDCOL (Infrastructure Development Company Ltd.) 236

268 Index

IGCP (International Gorilla Conservation Programme) 103
IGU see International Geographic Union
Impoverishment Risk and Reconstruction (IRR) model 186
income: food spending in relation to 167, 169–170, 172–173; gender equality/inequality in 34, 158–160, **159**, *159*; resettlement and restoration of 192, 196, 197; scarcity of (*see* poverty; scarcity and landlessness); *see also* socio-economic status
industry *see* business and industry
infrastructure: disaster and vulnerability research on 211–212, 213; displacement and resettlement in relation to 185–186, 191–192, 196–197; energy 20, 232; food security affected by **174**, 177; sanitation 20, 197; transportation 20, 212; urban strains on 20; water 20, 197, 212
Infrastructure Development Company Ltd. (IDCOL) 236
inheritance, women's rights to 33
Institute of British Geographers 30
Integrated Food Assisted Development Programme (FADEP) 40
Intergovernmental Panel on Climate Change (IPCC) 146
International Geographic Union (IGU): Commission on Commercial Activities 36; Commission on Gender 30; Population Commission 46–47
International Gorilla Conservation Programme (IGCP) 103
IPCC (Intergovernmental Panel on Climate Change) 146
IRR (Impoverishment Risk and Reconstruction) model 186
Islam, M. Aminul 35, 49
Islam, M. Shahidul 3, 97
Islam, Nazrul 2, 11, 18, 35, 49
Islam, Sheikh Tawhidul 3, 146, 254
Islam, women's rights in 33–34

Jahan, Akhter 34
Jahan, Rounaq 30
Jahangirnagar University 38, 49, 50, 256
Jamuna Bridge Access Road Project (JBARP) 185
Jamuna-Meghna River Erosion Mitigation Project (JMREMP) 185
Jamuna Multipurpose Bridge Project (JMBP) 4, 127, 185, 187, 190–192, 196–198
Japan International Coopery (JICA) 185
JBARP (Jamuna Bridge Access Road Project) 185
JICA (Japan International Coopery) 185
JMBP (Jamuna Multipurpose Bridge Project) 4, 127, 185, 187, 190–192, 196–198
JMREMP (Jamuna-Meghna River Erosion Mitigation Project) 185
Jones, Geoff 180

Kabir, Md. Humayun 4, 222
Kamaluddin, AFM 49
Karnafuli Fertilizer Co. Ltd. 127
Karnafuli Multipurpose Hydroelectric Project (KMHEP) 235
KB (Krishi Bank) 84
Kelley, Florence 44
Khan, Amanat Ullah 18
Khan, Fazle Karim 36, 48
Khan, Jafar Reza 49
Khatun, Hafiza 4, 35, 37, 38, 185
Khulna 12, 13, 115
KMHEP (Karnafuli Multipurpose Hydroelectric Project) 235
knowledge gaps, filling 256
Kosiniski, L. A. 47, *47*
Krishi Bank (KB) 84

lake sediment cores 100–101, *102*
land: agricultural 21, 74–93, 172–173, 180, 222, 224, 226–227, 229; capability of 226; displacement and resettlement 4, 49–50, 185–200; environmental assessments of 124, **128**, 132, **137–142**; gender and

access to 29; land-human ratio 11; natural resource appraisal of 222, 224, 225–228, 228, 238; ownership of 16, 19, 157, *157*, 186, 188, 197; pollution of 27, 225; price of 19; publicly owned 16, 186, 188; research and scholarship on 21; sale or loss of 87; scarcity and landlessness 2–3, 74–93, 157–158, 173; urbanization and distribution of 15, 19, 21, 173; *see also* land use classification and planning; soil
Land Acquisition Act 1894 (Act I of 1894) 187
Land Occupancy Survey (LOS) 75
landslides 108, 118–119, 176, 215
Land use Advisory Project 227
land use classification and planning: climate change crisis and 161; disaster and vulnerability research on 212; environmental impact assessments 127, 132, 135; food security and 173, 179; natural resource appraisal on 224, 225–226, 227, 238; urban 12, 15, 21
legal issues and rights: displacement and resettlement 186, 187–188, 198–199, 200; scarcity and land dispossession related to 89, 90; women's 29, 33, 197; *see also* crimes
lightning 151–152
literature reviews: definition and description of 241–243; ecological integrity understanding through 250–251; general 245; geography-specific 249–252; historical 245; historical context understanding through 251–252; literature searches for 247–248; literature sources for 246–247; location and context ascribed in 249–250; meta-analysis 246; methodological 245; overview of 4, 241, 252; scale understanding through 251; systematic 246; theoretical 245; traditional or narrative 245; usefulness and purpose of 243, **244–245**; writing of 248–249

lithostratigraphic analysis: deep sea cores in 99–100; field descriptions of sediments in 103–104, **104**; ice cores on 99; laboratory analysis of sediments in 104–105; lake sediment cores in 100–101, *102*; lithology 101–103; palaeogeographic 99–105
Livelihood Vulnerability Index (LVI) 215
loans and credit: displacement and access to 192, 196, 197, 199; food access through 170; gender and access to 29; scarcity management via 83–86, **85**
local trends, current state, and future of geography: future developments 257–259; geographical methods used 255; major contributions of discipline 255–256; overview of 4, 254
LOS (Land Occupancy Survey) 75
LVI (Livelihood Vulnerability Index) 215

Mahalanobis, P. C. 250–251
Malla, U. M. 36
mapping: current use of 255; disaster and vulnerability 210–216; environmental impact assessments 132, 135, 144; food insecurity 168, 173–177, 178–179; GIS for (*see* Geographical Information System); HIV/AIDS geography 55, 57, 63; literature reviews of 249–250, 252; natural resource appraisal 223, 225–226, 227–228, 230, 231, 238; population geography 44, 49, 50, 51, 52; poverty 44
marriage: scarcity and land dispossession in relation to 84, 86, 89, 92; women's rights in 33
Marxism, geography theory reflecting 1
MAT (Modern Analogue Technique) 98
MDGs (Millennium Development Goals) 178

metropolitan cities 13; *see also* urbanization
migration: gender research on 37; HIV/AIDS geography and 56–57; population geography study of 48–49; rural-urban 12, 16, 17, 173; socio-economic status after 17; urban growth from 12, 14, 16, 17
military administration 12
Millennium Development Goals (MDGs) 178
Ministry of Women and Children Affairs (MoWCA) 40, 161
Mitra, Satish Chandra 250
Modern Analogue Technique (MAT) 98
Momsen, Janet 30
Mondjanngni, A. 47
monsoons: climate change crisis and 150, **150**, 153, 160–161; palaeogeographic study of 97, 98, 100–101, *102*, 108, 113, 117–118; scarcity and vulnerability to 82, 84
MoWCA (Ministry of Women and Children Affairs) 40, 161
Multi-criteria Decision Analysis 215
Musa, AKM 180

National Food Policy (NFP) 167
National Land Use Policy (2001) 127
National Oceanographic Research Institute (NORI) 131–135, **133–134**
National Policy for Women's Advancement (NPWA) 40
National Policy on Resettlement and Rehabilitation 2007 (NPRR) 198–199, 200
National Water Policy (1999) 127
natural hazard analysis 211
natural resource appraisal: in Bangladesh 224–237; of energy 232–237, **234**, **236–237**; of forests 222, 224, **231**, 231–232; of land 222, 224, 225–228, **228**, 238; need for 223–224, 238; overview of 4, 222–223, 238–239; population pressure and 222, 223, 229, 237; process of 223–224;

resource demand in 223, 224, 238; resource inventory for 223–224, 227, 238; resource supply in 223–224, 238; of wetlands 222, 228–230
Nazem, Nurul Islam 18
NFP (National Food Policy) 167
Nizamuddin, K. 37
noise pollution, urban 13
non-renewable energy 232–234
NORI (National Oceanographic Research Institute) 131–135, **133–134**
NPRR (National Policy on Resettlement and Rehabilitation 2007) 198–199, 200
NPWA (National Policy for Women's Advancement) 40

oil and natural gas 232–234
Ordered Weighted Average (OWA) 215

Padma Multipurpose Bridge Project (PMBP) 185, 191, 192
palaeogeography and palaeoenvironment: of Bangladesh 112–120; dating of 108–109, **110–111**; deep sea cores on 99–100; diatom analysis of 106–108, *107*; field descriptions of sediments on 103–104, **104**; GBM delta formation in 112–113; ice cores in 99; laboratory analysis of sediments on 104–105; lake sediment cores on 100–101, *102*; landslides in 108, 118–119; lithology on 101–103; lithostratigraphic analysis of 99–105; monsoon characteristics in 97, 98, 100–101, *102*, 108, 113, 117–118; overview of 3, 97; peat accumulation in 120; pollen analysis of 98, 105–106, *107*, 115; sea level change in 3, 97, 98, 99, 105, 106, 108, 109, **113**, 113–114, *114*, 115; seismic activities in 118; signatures and techniques to reconstruct 98; vegetation history in 98, 99, 105, 109, 115–117, *116*

Index 271

PAR (Pressure and Release) model 208–209, 210
Paul, Alak 2, 48, 54, 254
Paul, Shitangsu Kumar 4, 202
peat, palaeogeographic study of 120
periphery geography 1, 4–6, 257–259; *see also* geography in Bangladesh
philosophy, geography theory reflecting 1
physical geography: climate change crisis 3, 49–50, 135–144, 146–161, 172, 206, 211; environmental impact assessments 3, 124–144; overview of 3; palaeogeography and palaeoenvironment 3, 97–120; research and scholarship in 5–6
plants *see* agriculture; flora and fauna
PMBP (Padma Multipurpose Bridge Project) 185, 191, 192
political context: Bangladeshi unrest in 34; disaster and vulnerability research consideration of 206, 209; gender geography in 30–31, 32; HIV/AIDS geography in 54, 57–58, 60, 62–63; population geography in 50; renewable energy in 234; scarcity and landlessness in 74, 89, 90, 92; urbanization in 13, 18
pollen analysis 98, 105–106, *107*, 115
population: Bangladeshi 222; global 222, 223; pressure of, on natural resources 222, 223, 229, 237; urban 11, 12, 13
population geography: background of 27; in Bangladesh 48–50; challenges and future directions in 50–51, 52; climate change crisis study in 49–50; conceptual and theoretical approaches to 45–46; definition of 45–46, 48; development linkage with 50–51; displacement and resettlement study in 49–50; epidemiologic methods in 44; evolution of 45–47; gender studies in 27; major theoretical framework for 48–49; methodological aspects of 46–47, 51; overview of 2, 44, 51–52; research and scholarship in 48–50
poverty: climate change crisis nexus with 156–158, *157*; development programmes increasing 28, 186–187; disasters and vulnerability with 215; displacement and resettlement leading to 186–187, 190, 200; food insecurity and 167, 170–171, 172–173, **174**, 177, 179; gender geography study of 28, 29, 31, 34; population geography and mapping of 44; scarcity and landlessness with 2–3, 74–93, 157–158, 173; urban 16–18, 21; women in 17, 28, 29, 31, 34
Poverty Reduction Strategy Paper (PRSP) 179
preparedness analysis 212
Pressure and Release (PAR) model 208–209, 210
prevention analysis 212
prostitution and sex work 34, 58, 59–60, 61, 62
Protection of Urban Wetlands and Open Spaces (2000) 127
PRSP (Poverty Reduction Strategy Paper) 179
public housing 18
publicly owned land 16, 186, 188

radiocarbon dating 109, **110–111**
Rahman, Bangabandhu Sheikh Mujibur 16
Rahman, M. Fazlur 180
Rajshahi 12, 13, 21
Rajshahi University 15, 49, 256
Ramsar Convention Bureau 229
Rasheed, K. B. Sajjadur 35, 37, 49
Regional Road Management Project (RRMP) 185
religion and gender 33–34
Remote Sensing 49, 132, 135, 144, 210, 226, 231, 255
renewable energy 232, **234**, 234–237, **236–237**
Renewable Energy Policy (REP) 236
research and scholarship: applied geography 5–6; current and

future 4, 254–258; disaster and vulnerability (*see* disaster and vulnerability research); food security 173–180; gender geography 34–39; HIV/AIDS geography 55–64; human geography 5–6; literature reviews 4, 241–252; physical geography 5–6; population geography 48–50; scarcity and landlessness 76–93; urbanization 11, 14–21
response analysis 212
risk: disaster and vulnerability research on 202, **203–205**, 205–210; displacement and resettlement mitigation of 186–187; HIV/AIDS geography on 56, 58, 61–62, 63; risk and hazard model on 207, 210
Ritter, Carl 26
riverbank erosion: disaster and vulnerability research related to 215; displacement and resettlement due to 197–198, 199, 200; research and scholarship on 20; scarcity and land loss due to 77, **78**, 82, 89, 90; women impacted by 34
RMP (Rural Maintenance Programme) 41
Rokeya Sakhawat Hossain, Begum 33, 34
RRMP (Regional Road Management Project) 185
rural geography: agriculture in (*see* agriculture); food security in 4, 83, 92, 161, 167–181; migration to 14; rural-urban migration 12, 16, 17, 173; scarcity and landlessness in 2–3, 74–93, 157–158, 173; sustainable development of 167, 179–180, 181n3; technology changes to 257; women in 34, 35, 39, 41
Rural Maintenance Programme (RMP) 41

Saleheen, Mesbah-us- 49
salinity intrusion 82, 151, 153–154, **155**

sanitation 20, 170, **174**, 177, 197
Sanyal, Nandini 4, 241
SAS (Statistical Analysis Systems) 178
scarcity and landlessness: coping strategies for 83; definition of 75; determinants of 90–92, *91*; disasters and vulnerability to 75, 76–77, **78–79**, 82–83, 89, 90, 92, 157–158; food insecurity with 83, 92, 173; generational 90; growth of 75–76; health concerns affecting 82, 83, 86, 89, 92; household asset sale and dispossession with 83–86, **85**; household-specific 82–83, 87; land sale or loss 87; loans to manage 83–86, **85**; overview of 2–3, 74–76; political context for 74, 89, 90, 92; reasons necessitating loans/sales with 86–90, **88**; research and scholarship in 76–93; rural scarcity in Bangladesh 77, **80**, 81–83; seasonal or temporal variations in 77, **80**, 81–83, 89–90, *91*, 91–92; social context for 74, 89, 92; sources of loans/credit 84–86, **85**; spatial context for 76, 81–83, 87, 89, *91*, 91–92
SCIAs (Strategic Cumulative Impact Assessments) 136, 144
SDGs (Sustainable Development Goals) 51, 179
sea level change: climate change crisis and 151, 155, 172; disaster and vulnerability research related to 215; palaeogeographic study of 3, 97, 98, 99, 105, 106, 108, 109, **113**, 113–114, *114*, 115; population geography study of 49
SEAs (Strategic Environmental Assessments) 126, 135–144, **137–142**, *143*
seismic activities *see* earthquakes
Sen, Gopal Chandra 180
SEVI (Spatial Multi-Criteria Social Vulnerability Index) 213
sexuality: HIV/AIDS and 58, 59–60, 61, 62; prostitution and sex work 34, 58, 59–60, 61, 62; women's liberation of 31

Shamsuddin, Dara 3, 124, 251–252
Shiva, Vandana 28
Shrivastava, V. K. 36
SIAs (Social Impact Assessments) 126
Sikder, Abdur Rashid 180
Silberstein, Brigitte 35
Singh, C. P. 36
slums: displacement and resettlement from 199, 200; scarcity and landlessness in 76; urbanization and 16–17, 18, 21
Snow, J. 44
social context: climate change crisis in 156–160, *157*, **159**, *159*; displacement, resettlement and changes to 186, 192, 196, 197, 199, 200; environmental impacts on, assessing 124, 126, 132, **137–142**; food security in 172–173; gender geography in 27, 29–31, 32, 37, 39–40, 41; HIV/AIDS distribution in 54–64; scarcity and landlessness in 74, 89, 92; social geography study of 27; *see also* socio-economic status
Social Impact Assessments (SIAs) 126
social theory, geography theory reflecting 1
Social Vulnerability Index (SoVI) 213
socio-economic status: climate change crisis nexus with 156–158, *157*; development programmes and 28, 186–187; disaster and vulnerability research consideration of 206, 208–210, 211, 213–215; gender geography study of 28, 29, 31, 34; HIV/AIDS geography in relation to 57, 58, 59, 62, 63; research and scholarship on 16; scarcity and landlessness affecting 74, 76, 92, 157–158; social geography on 27; technology influences on 257; urban 16–18, 21; women's 17, 28, 29, 31, 34; *see also* income; poverty
soil: contamination 27, 225; environmental assessments of 124, **128**, 132; health of, affecting food security 172, 180; natural resource appraisal of 227–228, **228**; salinity intrusion 82, 153–154, 155; *see also* land
solar energy 232, *234*, **234**, 236–237, **236–237**
Sorre, Maximilian 27
South-West Road Network Development Project (SRNDP) 185
SoVI (Social Vulnerability Index) 213
spatial and temporal context: climate change crisis in 146–147, 151–156; current use of 255; disaster and vulnerability research in 208, 209–210, 211, 213–216; displacement and resettlement in 4, 49–50, 185–200; food security in 171; gender geography in 27, 32–33; HIV/AIDS distribution in 55–56, 57; literature reviews on 249–250, 251–252; mapping of (*see* mapping); palaeogeography and palaeoenvironment 3, 97–120; population distribution in 44, 45–46; scarcity and landlessness in 76, 77, **80**, 81–83, 87, 89–90, *91*, 91–92; urban 16
Spatial Multi-Criteria Analysis 213–216
Spatial Multi-Criteria Social Vulnerability Index (SEVI) 213
SPSS (Statistical Package for the Social Sciences) 51, 178
squatters settlements 16–17, 21, 197
SRNDP (South-West Road Network Development Project) 185
Stamp, Dudley 225, 226
STATA 51
Statistical Analysis Systems (SAS) 178
Statistical Package for the Social Sciences (SPSS) 51, 178
Strategic Cumulative Impact Assessments (SCIAs) 136, 144
Strategic Environmental Assessments (SEAs) 126, 135–144, **137–142**, *143*
Sultana, Nayar 34
sustainable development: agricultural 167, 179, 180, 180n1; defined 181n3; food security in context of 4, 167–168, 171–173, 178–180; population

geography and 50, 51, 52; rural 167, 179–180, 180–181n2; Sustainable Development Goals 51, 179; women's importance to 39

Taha, Abu 49
TBDLRP (Tongi-Bhairab Double Line Railway Project) 185
temperature: agricultural effects of 81, 176; climate change crisis and 147–150, **148–149**, 151, 161; palaeo- 98, 118
thunderstorms 151–152
tidal energy 234
Tongi-Bhairab Double Line Railway Project (TBDLRP) 185
Toponyms of Bangladesh: Footprints of History (Shamsuddin) 251–252
trade 12–13, 19
transportation: disaster and vulnerability research on 212; employment in 19, 157; HIV/AIDS among workers in 59, 60, 62; infrastructure for 20, 212; research and scholarship on 15, 20; urbanization and access to 13, 20
Trewartha, G. T. 27, 45, 46
tsunamis 118

UDHR (Universal Declaration of Human Rights) 168
UNICEF 17–18
United Nations: Conference on Environment and Development 29; Decade for Women 28, 30; demographic data publication by 46; development programmes 28–29, 227; Framework Convention on Climate Change 146; Intergovernmental Panel on Climate Change 146; UNICEF 17–18; Universal Declaration of Human Rights 168; World Food Programme 161, 172, 173, 175, 179
Universal Declaration of Human Rights (UDHR) 168
University of Dhaka *see* Dhaka University
Upazilas 13, 14, 173–175

urbanization: Dhaka's primacy and 12–13, 16, 21; economy and 12, 14, 19, 21; employment and 19, 20–21; environmental concerns with 13, 19–20; food security affected by 173; geographic perspective on 2, 11–21; housing challenges with 16–17, 18–19, 21; infrastructure limitations with 20; land access and distribution with 15, 19, 21, 173; metropolitanization and 13; patterns and processes of 11–12, 16; post-liberation research on 15–21; poverty and socio-economic status with 16–18, 21; pre-liberation research on 14–15; rate of 12; reclassification of urban boundaries 12; research and scholarship on 11, 14–21; rural-urban migration 12, 16, 17, 173; secondary cities and towns 14; slums and squatter settlements with 16–17, 18, 21; transportation and 13, 20; urban population statistics 11, 12, 13

vegetation *see* agriculture; flora and fauna
VGDUP (Vulnerable Group Development for Ultra Poor) 40, 161
VGF (Vulnerable Group Feeding) 161
Vilal de la Blache, Paul 27
violence against women 29
von Post, Lennart 105
voting rights, women's 34
vulnerability research *see* disaster and vulnerability research
Vulnerable Group Development for Ultra Poor (VGDUP) 40, 161
Vulnerable Group Feeding (VGF) 161

Wallace, B. 34, 35
War of Liberation 16, 40
WARPO (Water Resources Planning Organization) 125, 126
water: droughts from lack of 82, 92, 151, 153, 171, 176; environmental assessments of 124, 125, **129**,

132, **137–142**; floods (*see* floods); hydropower 232, 234, **234**, 235; infrastructure for 20, 197, 212; natural resource appraisal of 224; pollution 13, 27, 229; riverbank erosion (*see* riverbank erosion); salinity intrusion 82, 151, 153–154, **155**; sea level (*see* sea level change); tidal energy 234; urban concerns with 13, 19–20; wetlands 127, 222, 228–230
Water Resources Planning Organization (WARPO) 125, 126
WCED (World Commission on Environment and Development) 181n3
WebDMAP 57
Weighted Liner Combination (WLC) 215
Western geography 1, 4–5, 6, 257
wetlands 127, 222, 228–230
White, Gilbert F. 206
WID (Women in Development) 30, 31, 40
Wildlife Conservation and Security Act (1992) 127
wind energy 234, **234**, 235
WLC (Weighted Liner Combination) 215
women: academic study of 30–31; Bangladesh as context for 33–39; climate change crisis effects on 158–160, **159**, *159*; Decade for Women 28, 30; development programmes and 28, 29–30, 39–41; displacement and resettlement of 192, 197, 198, 199; education for 33; employment of 20–21, 29, 34, 39–40, 41, 158–160, **159**, *159*; environmental assessments of effects on **130**; environmental concerns addressed by 28–29; equality for 29–30, 31; feminist advocacy and epistemology for 29–32; food security/insecurity for 40–41, 170, 171; gender geography on 2, 26–41, 58, 59–60, 158–160, 170, 171; HIV/AIDS among 58, 59–60, 61, 62; legal rights of 29, 33, 197; migration by 17, 37; poverty and socio-economic status of 17, 28, 29, 31, 34; social context for 27, 29–31, 32, 37, 39–40, 41; spatial context for issues of 27, 32–33; urban context for 17, 20–21; violence against 29
Women, Work and Environment: Studies in Gender Geography (Ahsan, Ahmad, Eusuf & Nizamuddin) 37
Women and Development course 36
Women Environment and Development 28
Women in Development (WID) 30, 31, 40
Women's Conferences 28–29
Women's Liberation Movement 28, 29
World Bank: development programmes 185, 186–187; environmental impact assessments 125–126, 127; salinity intrusion study for 153; social vulnerability assessment 214
World Commission on Environment and Development (WCED) 181n3
World Food Programme 161, 172, 173, 175, 179
World Food Summit 168
World Women Congress for Healthy Planet 29

Zelinsky, W. 45–46

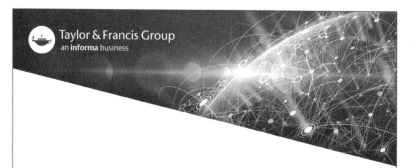